T0075105

Unravelling Long COVID

Your book purchase provides you with free access to unravellinglongcovid.com

Long-Covid is an evolving medical disorder, and it is essential that patients and their healthcare providers receive updated, important information in a timely fashion. To accomplish this, we have linked this book with our new website, unravellinglongcovid.com. This provides a unique opportunity for the reader to learn about the latest medical and scientific studies on Long-Covid. This will begin with a review of any significant information that we have learned during the months that this book was in publication. The website will be updated at bi-weekly intervals and sooner if an essential new study is reported. It will also offer the reader an opportunity to ask questions of the book's authors.

Don Goldenberg, MD

Unravelling Long COVID

Don Goldenberg, MD
Emeritus Professor of Medicine
Tufts University School of Medicine
Boston, MA, USA
Adjunct Faculty, Departments of Medicine, Nursing
Oregon Health Sciences University
Portland, OR, USA

Marc Dichter, MD, PhD
Emeritus Professor of Neurology
University of Pennsylvania School of Medicine
Philadelphia, PA, USA

WILEY Blackwell

This edition first published 2023

Registered Offices
John Wiley & Sons, Inc., 111 River Street, Hoboken, NJ 07030, USA
John Wiley & Sons Ltd, The Atrium, Southern Gate, Chichester, West Sussex, PO19 8SQ, UK

For details of our global editorial offices, customer services, and more information about Wiley products visit us at www.wiley.com.

Wiley also publishes its books in a variety of electronic formats and by print-on-demand. Some content that appears in standard print versions of this book may not be available in other formats.

Library of Congress Cataloging-in-Publication Data applied for
ISBN: 9781119891307 (hardback)

Cover Design: Wiley
Cover Image: © Ralwell/Shutterstock

Set in 9.5/12.5pt STIXTwoText by Straive, Pondicherry, India

SKY10037657_110122

Contents

Introduction

As the COVID-19 pandemic stretched on, it became abundantly clear that many patients had persistent symptoms long after all signs of the initial infection vanished. These lingering symptoms persist for months in 30% to 80% of patients who were hospitalized with COVID-19 infection and 10% to 30% who were not hospitalized. They appear in people across a wide spectrum of COVID-19 sufferers, including those with mild illness or even asymptomatic infections. These long-lasting symptoms are present in adults and children, and current estimates suggest that at least 25 million Americans, and ten times that worldwide, have been or will be affected. Many experts believe that the persistent symptoms following acute COVID-19 will become the next major, global public health disaster.

Understanding these persistent symptoms has been wrought with confusion. Researchers and clinicians have not even agreed upon a name, with long COVID or long-haulers' syndrome often used and post-acute sequelae of SARS-Cov-2 (PASC) recommended by most medical societies. There is not a uniform definition, and the proposed diagnostic criteria focus on the duration of symptoms rather than the nature of the symptoms. At the present time, patients, healthcare professionals, and scientists find themselves looking for answers about the nature of long COVID.

We believe that there are two main issues that have interfered with understanding long COVID. The first involves a failure to distinguish patients with an obvious source for their persistent symptoms from the many patients whose symptoms cannot be easily explained. The term long COVID was coined by patients to describe a constellation of persistent symptoms that were not being adequately acknowledged nor explained. Patients saw multiple physicians and were undergoing numerous tests, without answers.

We have made such a distinction, outlined initially in Section 1, by grouping patients with organ damage after COVID-19, what we term **long-COVID disease**, apart from those whose persistent symptoms are unexplained, what we call **long-COVID syndrome**. This grouping answers a diagnostic dilemma, described by Alwan in the July 2021 *Science*, "One important issue is whether 'Long Covid', as a label, will include organ pathology diagnosed weeks or months after COVID-19, or whether these cases move out into an alternative diagnostic category, leaving only those with 'unexplained' symptoms as having Long Covid" [1].

In the book's first section, we detail the persistent symptoms and clinical course of patients with organ damage, suggesting that their disease pathways and outcomes are similar to those of patients after any severe illness. The perplexing issue is how to better understand the lingering symptoms following COVID-19 infection that are not explained by organ damage. These persistent symptoms have been called "medicine's blind spot." Such common, persistent symptoms have been a source of confusion and controversy for centuries. In Section 2, we compare such persistent symptoms to those in the general population and then in medical conditions most often compared to long COVID, including chronic fatigue syndrome/myalgic encephalomyelitis (CFS/ME) and fibromyalgia.

In Section 3, we explore the mechanisms underlying long-COVID syndrome. This requires an appreciation of brain homeostasis in health and disease and how alterations in central nervous system pathways can explain these symptoms. We suggest that, rather than a traditional autoimmune disease, long-COVID syndrome fits best within a neuroimmunologic framework.

In the final section of the book, we discuss current and future patient evaluation, including innovative research in long COVID. Finally, we discuss ongoing treatment programs, including dedicated long-COVID clinics throughout the world and guidelines for primary care awareness and optimal management. We examine the important role of patient advocacy and the potential impact of consumer-directed research. We suggest new approaches to balance physician and patient perspectives.

One of us, DG, was already researching long COVID, while writing his book *Covid's Impact on Heath and Health Care Workers*. DG, Emeritus Professor of Medicine at Tufts University School of Medicine, is a rheumatologist and an international expert in many of the illnesses often compared to long COVID, including CFS/ME and fibromyalgia. He has evaluated and treated more than twenty-thousand patients with these conditions and has experienced the frustrations of his profession's unsuccessful attempts to understand the causes of these very disabling conditions that inflict suffering on so many people. His interests in long COVID were stimulated by the common symptoms affecting long-COVID patients to those he has dealt with in patients for four decades.

Simultaneously and independently, his long-time friend and medical colleague, MD, Emeritus Professor of Neurology and Former Director of the Mahoney Institute of Neurological Sciences at the Perelman School of Medicine, University of Pennsylvania, is a neurologist physician-scientist with a life-long research interest in the intricate workings of the brain under normal conditions and in a variety of disease states. MD is particularly interested in researching and understanding the persistent or new brain-related symptoms that occur in people recovering from COVID-19. Despite living on opposite sides of the United States from one another, Portland, Oregon and Philadelphia, Pennsylvania, the two specialists decided to write this book together, meshing their complementary clinical and research interests.

The authors' backgrounds in medicine are ideally suited to help people understand this vexing and mysterious disorder. DG has spent much of his career dealing with chronic illnesses that cannot be easily pigeonholed as physical or psychologic and are best appreciated from a biopsychological illness model. MD has focused his research on brain disease from a biomedical disease model. We believe that long COVID can be best understood by integrating biomedical and biopsychological illness models.

Reference

1 Alwan, N.A. (2021). The road to addressing Long Covid. *Science* 373 (6554): 491–493. https://doi.org/10.1126/science.abg7113.

Section 1

Long-COVID Disease

1

Long-COVID Disease or Long-COVID Syndrome?

Defining Long COVID

What does long COVID mean? In an all-encompassing fashion, it refers to any symptoms following a SARS-CoV-2 infection that persist for an extended time. It is not uncommon for symptoms to persist after an infection. Long COVID is a new term, introduced by patients, to account for multiple symptoms that last months and interfere with daily life, yet have no clear medical explanation.

Initially, definitions of long COVID were based primarily on the duration of symptoms, with symptoms lasting for more than three months considered unusual. Since most individuals with SARS-CoV-2 infection recover completely within three months, we adopted the time frame for long COVID to include symptoms that last more than three months [1]. Subsequent case definitions included the most prominent lingering symptoms. Fatigue, shortness of breath (dyspnea), musculoskeletal pain, cognitive disturbances, sleep and mood disturbances, and headaches are the most common persistent symptoms; we included these symptoms in our long-COVID definition (Table 1.1).

What has made long COVID so important and controversial is how these characteristic symptoms persisted long after all signs of the initial infection disappeared. The National Institutes of Health (NIH) suggested the term, post-acute sequelae of SARS-Cov-2 (PASC) and defined post-acute symptoms as those that develop during or after COVID-19 infection that cannot be attributed to an alternative diagnosis. The National Institute for Health and Care Excellence (NICE), Scottish Intercollegiate Guidelines Network (SIGN), and Royal College of General Physicians termed the symptoms post-COVID syndrome, defined as, "Signs and symptoms that develop during or after an infection consistent with COVID-19, continue for more than 12 weeks and are not explained by an alternative diagnosis. It usually presents with clusters of symptoms, often overlapping, which can

Unravelling Long COVID, First Edition. Don Goldenberg and Marc Dichter.

Table 1.1 Our definition of long COVID.

Documented or Suspected SARS-Cov-2 infection.
Duration of symptoms greater than three months.
More than three of the following symptoms:
Fatigue
Dyspnea
Musculoskeletal pain
Headaches
Cognitive disturbances
Sleep disturbances
Mood disturbances

fluctuate and change over time and can affect any system in the body. Post-COVID-19 syndrome may be considered before 12 weeks while the possibility of an alternative underlying disease is also being assessed." [1, 2].

How do we define the absence of an underlying disease? For patients admitted to an intensive care unit (ICU) with severe COVID, these persistent symptoms align with a phenomenon often called post-ICU syndrome. Dr. Anthony David, Professor at the Institute of Mental Health, University College, London, stated in December 2021, "If a patient recovers from the acute respiratory illness, but remains short of breath and is found to have pulmonary fibrosis or pericarditis by accepted criteria, or, experiences brain fog and mental slowing, later linked to microvascular infarcts on magnetic resonance imaging (MRI)—can they be removed from the post-COVID-19 cohort? I would say yes. Their condition may be unusual, and it may be serious, but it is not mysterious. These conditions add to the tally of morbidity caused by COVID-19, but not to post-COVID-19 syndrome [3]." For those patients with organ damage during the initial infection, we will use the term long-COVID disease. When the persistent symptoms following a SARS-Cov-2 infection remain unexplained we use the term long-COVID syndrome.

Long-COVID Disease or Syndrome?

We believe that it is essential to recognize that long COVID is a disease in some situations whereas in others it is a syndrome. Disease is defined by organ damage, such as when a biopsy reveals cancer. A disease is characterized by its symptoms, such as pain or exhaustion, as well as physical signs, such as fever or swelling. In contrast, the term syndrome is applied to a medical disorder without obvious organ damage. Syndromes are diagnosed based solely by their symptoms. A syndrome is like a temporary placeholder for an illness, that may graduate to the more objective realm of a disease. Oftentimes, diseases originally considered syndromes were found to have a specific cause and/or organ damage/dysfunction, which lead to their recategorization as diseases.

Many patients have clinical evidence of organ damage during acute COVID infection. Their long-COVID symptoms follow the script of other severe, infectious diseases. The persistent symptoms correlate with the severity and duration of the acute infection. Almost all hospitalized patients have lung disease, and their persistent shortness of breath is the result of organ damage that may or may not be reversible. There is nothing mysterious about their long-lasting dyspnea and it can be measured by objective pulmonary abnormalities, such as pulmonary function tests and lung imaging. This is long-COVID disease.

However, other patients with persistent symptoms after a COVID-19 infection lack obvious organ damage and the underlying pathophysiologic mechanisms are unclear. Their physical examination, blood tests, X-rays, and imaging studies are normal. Syndromes are characterized by symptoms that involve many systems (multisystemic), occur together (cluster) and fluctuate in severity. These patients should be diagnosed with long-COVID syndrome, distinct from those patients suffering from well-described disease pathology.

To illustrate these differences, we will present two cases, one that we identify as long-COVID disease and the second, long-COVID syndrome.

Case 1. James, a 62-year-old man, was admitted to the hospital on June 1, 2020, because of increasing shortness of breath. His past medical history included adult-onset diabetes and obesity. He had experienced a cough and low-grade fever for three days, and a nasal swab tested positive for SARS-CoV-2 by polymerase chain reaction (PCR) the day before admission. Upon admission, he had a fever of 103° and was breathing rapidly. His initial chest X-ray demonstrated ground-glass opacities in both lungs and his oxygen saturation was 88%, normal oxygen saturation is greater than 94%. Over the first 48 hours, his breathing worsened despite nasal oxygen and prone positioning. He was transferred to the ICU where he was intubated and sedated as needed for mechanical ventilation. His treatment included corticosteroids and monoclonal antibodies in addition to the mechanical ventilation. After two weeks, the breathing tube was disconnected, and he was transferred from the ICU to a rehabilitation unit where he spent the next four weeks. In the rehabilitation unit, he needed a wheelchair at first, then graduated to a walker, but he was still profoundly weak. He described a constant worry about himself and his family, feeling "like being in a dark tunnel, trapped, and alone. I haven't seen my wife and children for more than a month, except on Zoom calls."

When James finally returned home, he was unable to stand without assistance and could only walk one block. He had lost 40 pounds. A repeat chest X-ray, taken one month after discharge, revealed scarring consistent with pulmonary fibrosis. Over the next six months, he continued to have shortness of breath with minimal activity despite an intensive course of pulmonary and occupational rehabilitation. Gradually, his ability to take care of himself and his pulmonary function tests improved slightly. One year after his hospitalization he said, "I'm still quickly exhausted. Even having my grandchildren over for a few hours is so draining. I worry that I will never get back to the way I was."

Case 2. Sarah, a 48-year-old female, began having symptoms that she suspected were related to a COVID-19 infection in March of 2020. She had been in good health with no chronic medical problems other than a long history of migraine headaches. Her acute symptoms included a low-grade fever, cough, headaches, and generalized muscle aches. Coronavirus testing was not yet widely available, but her primary care physician told her that she likely had COVID-19 and told her to self-quarantine for two weeks. Gradually, she felt better and returned to work as a nurse's aide but almost immediately stopped working, because "I was completely exhausted, mentally and physically. I was unable to do the simplest tasks. My heart kept racing and each time I tried to take a short walk, I had to stop and catch my breath. I found myself falling asleep throughout the day but then unable to sleep at night. The worst is this brain fog. I can't focus or concentrate on anything." Her primary care doctor examined her, ordered a chest X-ray and blood tests, but found no abnormalities. During the next few months, she saw a cardiologist, neurologist, and pulmonologist. A chest CT scan, pulmonary function tests, an MRI of the brain, and an echocardiogram were all normal.

The migraine headaches worsened, adding to her sense of dread. Her neurologist increased the dose of her migraine medications and thought she was becoming depressed. An antidepressant was prescribed but it did not help her mood or her exhaustion. The cardiologist was quite certain that her heart was not the problem, but he recommended further cardiac testing, including a cardiac MRI and coronary angiogram. These were also normal. Sarah was then referred to an endocrinologist and an immunologist. Their test results were also normal and could not make a specific diagnosis.

Now, 18 months after her initial infection, Sarah continues to feel short of breath and exhausted. "Before COVID, I worked out every day and I had run two marathons. Now, if I try to do even modest exercise I have to lay down for an hour. My chest hurts a lot, and it sometimes hurts just to take a breath. Even though all my heart tests are normal, I worry that they are missing something. I still can't concentrate, can't even write an email back to my friends. Some days I feel pretty good and then for no reason, all my symptoms get worse, and I am back to square one."

The first case represents what we call long-COVID disease, and James' persistent symptoms are readily explained by the severe lung infection that resulted in irreversible lung disease and the subsequent long rehabilitation. His subsequent

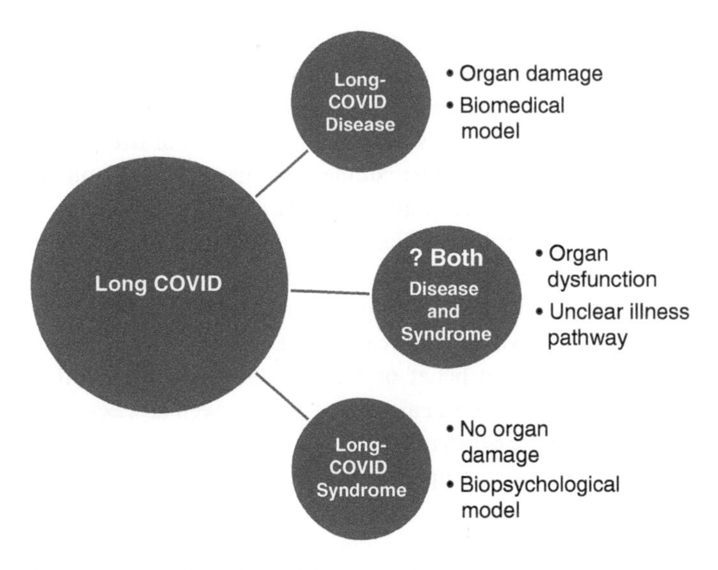

Figure 1.1 Defining long COVID as both a disease and a syndrome.

medical course and prolonged disability follow biomedical illness models (Figure 1.1). Sarah's symptoms are characteristic of long-COVID syndrome. Her persistent symptoms do not correlate with organ damage. Her protracted suffering may be better understood from a biopsychological illness model (Figure 1.1).

Unfortunately, most reports of long COVID did not record the presence or absence of organ damage. It is very likely that any patient hospitalized from acute SARS-CoV-2 infection had significant pulmonary disease. Therefore, we use hospitalization as a proxy for organ damage. Contrasting a series of hospitalized versus non-hospitalized patients provides us with a surrogate to compare long-COVID disease to long-COVID syndrome.

Regarding the documentation of a SARS-CoV-2 infection, we assumed that each hospitalized patient was a definite case of COVID-19. In non-hospitalized patients we used a positive laboratory test for acute SARS-CoV-2 to confirm the diagnosis of COVID-19.

We will focus on studies that include a control group of non-COVID-19 subjects. Many reports of long COVID failed to compare the prevalence of symptoms, organ damage, and outcome following COVID-19 to that of people in the general population or to patients with other diseases. This requires a controlled study [4]. This is especially important when evaluating symptoms that are common in the general population. At first glance, it may seem important if a hypothetical study found that 30% of non-hospitalized patients had headaches six months after COVID-19. However, the significance of that finding pales if 25% of the general population also had chronic headaches.

We recognize that our focus on cases with a confirmed SARS-CoV-2 infection and studies that had a control or comparison group limits some of our observations. Many patients with long COVID were never tested or may have had a false negative test, and confirmatory test results should not be required in the clinical care of patients with long COVID. However, studies requiring a confirmatory diagnosis of SARS-Cov-2 and using control populations are necessary for research. Such reports are necessary to avoid inherent bias in interpreting the findings. For example, in one study, self-reported long COVID was four times more common than a laboratory-confirmed diagnosis and the persistent symptoms differed in the self-reported cases from those in confirmed cases [5]. Long-COVID symptoms at one year correlated more with the belief of having COVID-19 than with a laboratory-confirmed diagnosis. Persistent pain and fatigue as well as sleep, cognitive, gastrointestinal, and mood disturbances were associated with a belief of having COVID-19 whereas the only symptom associated with a confirmatory SARS-CoV-2 test was a loss of smell.

We believe that distinguishing long-COVID disease from long-COVID syndrome is an important first step to understanding long COVID. However, many patients with prolonged symptoms do not fit neatly into either the disease or syndrome category. We think of them as having components of both a disease and a syndrome (Figure 1.1) James had severe lung disease and fit into our long-COVID disease category. His subsequent course can be explained by the ravages of that lung damage. Yet, many of his persistent symptoms, such as mood and cognitive disturbances, are likely related to his months in the hospital, social isolation, and sense of hopelessness. These are not direct effects from the lung damage. Sarah fits the biopsychological model of long-COVID syndrome, as manifested by the paucity of any abnormal tests nor evidence for organ pathology. However, as we will discuss in Section 3, the boundary between long-COVID disease and long-COVID syndrome becomes blurred when examining the brain (Figure 1.1). Neuroplasticity and neuroimmune mechanisms allow us to better understand these complex interactions.

Long COVID in Hospitalized Patients

Well-controlled studies have demonstrated that patients hospitalized with COVID-19 have much greater, persistent health problems than uninfected subjects. In a large study from the United Kingdom (UK), 48 000 patients hospitalized with COVID-19 infection were followed up at a mean of 140 days post-discharge and compared to subjects from the general population who did not have COVID-19 [6]. The two groups were matched for age, sex, ethnicity, co-morbidities, and body mass index. Nearly one-third of the COVID-19 patients had been readmitted and 10% had died after the initial hospitalization. Rates for readmission in those 140 days were fourfold greater than in non-COVID-19 subjects and rates of deaths were eightfold greater. During those four and one-half months, there were greater pulmonary symptoms and more cardiovascular, diabetes, kidney, and liver disorders in the COVID-19 patients compared to the non-infected controls.

More than 1200 COVID-19 patients discharged from a single hospital in Wuhan, China were followed up at 3, 6, and 12 months after their hospitalization [7]. These long-COVID patients were matched to community controls who did not have COVID-19. At one year, the COVID-19 patients had more overall medical symptoms, including more pain, problems with mobility, and mood disturbances, than did community controls. The median age of the long-COVID group was 59 years and 53% were men. At 186 days after hospitalization, more than 80% of the patients reported at least one symptom consistent with long COVID. There were significantly more persistent medical symptoms in these patients compared to the community controls who were not infected, 66% versus 33% (Table 1.2) [7]. At least one long-COVID symptom was present in 68% of the hospitalized group at 6 months and 49% at 12 months. There was a slight increase in dyspnea from 26% at 6 months to 30% at 12 months and in depression or anxiety, from 23% to 26% at 12 months.

Women had more long-COVID symptoms than men, including greater fatigue, anxiety, or depression, and greater dyspnea, which was documented with abnormal pulmonary function testing. There was no correlation of the fatigue, sleep disturbances, hair loss, smell disturbances, palpitations, or joint pain with the severity of respiratory difficulty during the hospitalization, which was graded as

Table 1.2 Symptoms at 6 and 12 months.[a]

Symptom (%)	6 months	12 months
At least one long-COVID symptom	68	49
Fatigue	52	20
Pain	27	29
Mood disturbances	23	26
Sleep disturbance	27	17
Joint pain	11	12
Palpitations/tachycardia	10	9

[a] Each symptom was significantly more common in COVID patients than controls.
Source: Based on Huang et al. [7].

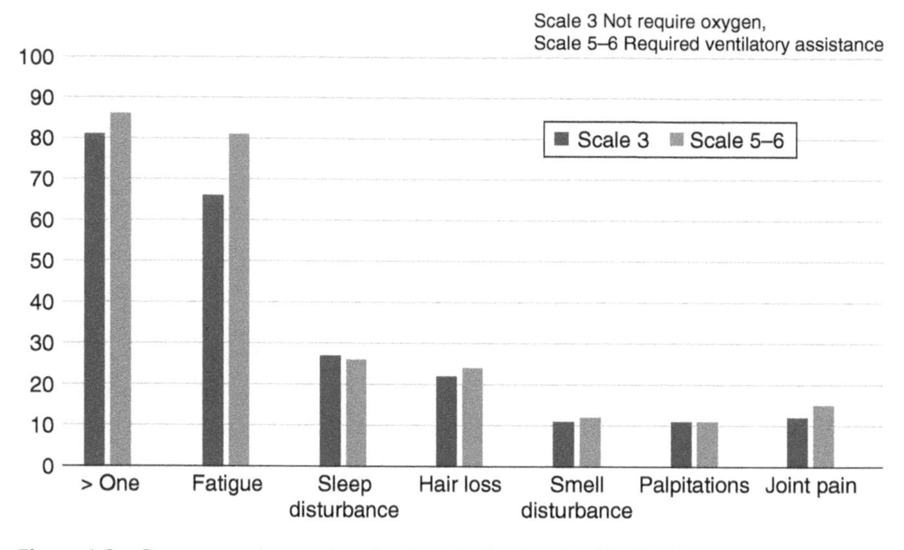

Figure 1.2 Symptoms six months after hospitalization for COVID-19.

Scale 3, not requiring supplemental oxygen versus Scale 5, requiring ventilatory assistance (Figure 1.2).

In controlled reports comparing hospitalized COVID-19 patients to non-infected community controls, about 10% of patients meet criteria for long COVID at three to six months after hospital discharge [6–8]. These differences, when comparing long-COVID symptoms in hospitalized patients to those in matched controls, tend to decrease over time but are still substantial at one year [7, 8].

A one-year, controlled study from Spain compared patients hospitalized with COVID-19 from March 1 to April 15, 2020, to patients hospitalized for another reason during that same time frame [9]. At least one long-COVID symptom was present at 12 months in 36% of the COVID-19 patients and 35% of the controls (Table 1.3). The only persistent symptoms at one year that were more common in the COVID-19 patients were upper respiratory symptoms, such as sore throat, cough or dysphonia, confusion or memory loss, and anxiety. This study confirms the importance of using a control population and the authors concluded that "These findings suggest that, rather than attributing persistent symptoms to COVID-19, it is the need for hospitalisation that prolongs long-term symptomatology."

In uncontrolled studies, long COVID has been present in most hospitalized patients. In one report of 9751 hospitalized COVID-19 patients, 73% had at least one persistent symptom 60 days after infection [10]. The most frequent symptoms were fatigue in 40%, dyspnea in 36%, sleep disturbances in 30%, memory loss in 27%, and anosmia in 20% with depression, anxiety, cognitive disturbance, palpitations, myalgias, and headache each about 15%. The PHOSP-COVID survey, a

Table 1.3 Long-COVID symptoms at one year.

Long-COVID symptom	% in COVID-19 patients	% in controls
Any long-COVID symptom	36	35
Fatigue	8	12
Muscle/joint pain	9	11
Dyspnea	15	12
Chest pain	1	2
Headache	3	3
Anxiety	7[a]	4
Confusion, memory loss	4[a]	2
Depression	5	4
Sleep disturbances	4	3
Upper respiratory symptoms	3.5[a]	1

[a] Significant increase in long-COVID patients versus controls.
Source: Modified from Rivera-Izquierdo et al. [9].

prospective, longitudinal study of more than 2000 COVID-19 patients discharged from UK hospitals in 2020–2021, found that 90% of hospitalized COVID-19 patients had at least one persistent symptom one year later, with an average of nine symptoms [11]. The most common persistent symptoms at one year were fatigue (60%), muscle pain (55%), sleep and cognitive disturbances (50%), and shortness of breath (50%). Persistent symptoms were associated with female sex, more severe acute illness, and co-morbidities, including obesity. Patients who required mechanical ventilation during their hospitalization and had multiple symptoms were less likely to fully recover at one year. Inflammatory laboratory markers, such as an elevated C-reactive protein (CRP), correlated with the severity of initial infection and poor outcome. Less than one-third of these patients described themselves as fully recovered one year after hospitalization. In contrast, in the report from Wuhan, China, 88% had returned to their previous work [7]. Of the 12% unable to return to their prior jobs, one-third noted that it was because of physical limitations after COVID-19.

Long COVID in Non-hospitalized Patients

In two, large, controlled studies from the UK, symptoms consistent with long COVID were sixfold more common in seropositive subjects compared to non-infected subjects [12, 13]. In the first, a general population survey, 3% of the

SARS-CoV-2 positive patients had at least one symptom consistent with long COVID at three months, compared to 0.5% in the controls. Approximately 5% of seropositive subjects had at least one symptom of long COVID at 12–16 weeks and 4.2% at 16–20 and 20–24 weeks. These were each statistically more prevalent than in seronegative controls.

Another large UK prospective survey, part of a mobile application termed "The COVID Symptom Study", launched on March 24, 2020 [13]. Subsequently, more than five million people registered for the app. An initial report of 4182 individuals who tested positive for SARS-CoV-2 found that 14% of participants had medical symptoms more than 28 days, 5% at more than 8 weeks, and 2% more than 12 weeks after infection. The most prominent symptoms that persisted for more than one month were fatigue, headache, dyspnea, and anosmia. These persistent symptoms correlated with increasing age, body mass index, the female sex, and having more symptoms at initial presentation. The five symptoms in the first week that were most predictive of long COVID were fatigue, headache, dyspnea, a hoarse voice, and myalgias.

In the REACT-2 study of more than 1.5 million UK residents, the weighted (controlled for the general population) prevalence of at least three long COVID symptoms at three months was 2.2%, including at least one-third that reported "a significant effect on my daily life" [14]. Long COVID was more common in women. This study included a small number of hospitalized patients.

Two studies compared persistent symptoms consistent with long COVID in infected healthcare workers (HCWs) to those in uninfected HCWs. In the first study, 140 HCWs were identified as SARS-CoV-2-positive cases with mild to moderate symptoms and compared to 1160 uninfected controls [15]. The control subjects remained asymptomatic and had negative PCR and antibody tests for SARS-CoV-2. Some symptoms, particularly a loss of taste and smell, were much more prominent in COVID-19 patients compared to controls. Nevertheless, many other symptoms, including mood and sleep disturbances, gastrointestinal disturbances, and hair loss, were just as common in SARS-CoV-2-negative as SARS-CoV-2-positive subjects.

The second study evaluated 2000 HCWs in Sweden at two, four, and eight months after what were considered mild SARS-CoV-2 infections and compared those symptoms to HCWs who had not been infected [16]. Previous SARS-CoV-2 infection status was based on the detection of a SARS-CoV-2 IgG antibody. The analysis included 323 seropositive and 1072 seronegative individuals of a similar age (median 45 years) that were 84% female. At least one or more symptoms were fourfold more common in seropositive individuals, with a loss of smell and/or taste, fatigue, and dyspnea all being significantly more prevalent. At two to four months, anosmia, ageusia, and fatigue were the most common symptoms, and each were significantly more common in seropositive patients. A loss of taste and smell were more than 30-fold greater in SARS-CoV-2-positive patients than in the controls.

As discussed above in hospitalized patients, the importance of control groups cannot be overstated. In the UK Symptoms Survey, 86% of PCR-positive subjects reported persistent fatigue but so did 58% of uninfected subjects; 77% of infected subjects had persistent headaches but so did 58% of PCR-negative individuals [13]. In contrast, a persistent loss of smell was present in 60% of the infected patients but only 7% of controls. In a report from the US, 90% of SARS-CoV-2-positive subjects were symptomatic three months after infection with an average of nine symptoms but so were 60% of the seronegative subjects with an average of five persistent symptoms [17].

However, reports of long COVID from online support groups have been invaluable in providing more details about persistent symptoms and their impact on daily activities. For example, using patients' recall, we can compare more than 25 symptoms during the SARS-CoV-2 infection to those at both three- and six-months post-infection [18, 19] (Figure 1.3).

As noted in controlled studies, fatigue is the most persistent symptom in long COVID, present in 90% of patients during the acute infection as well as in 80% at three and six months. In contrast, a loss of taste and smell, present in more than

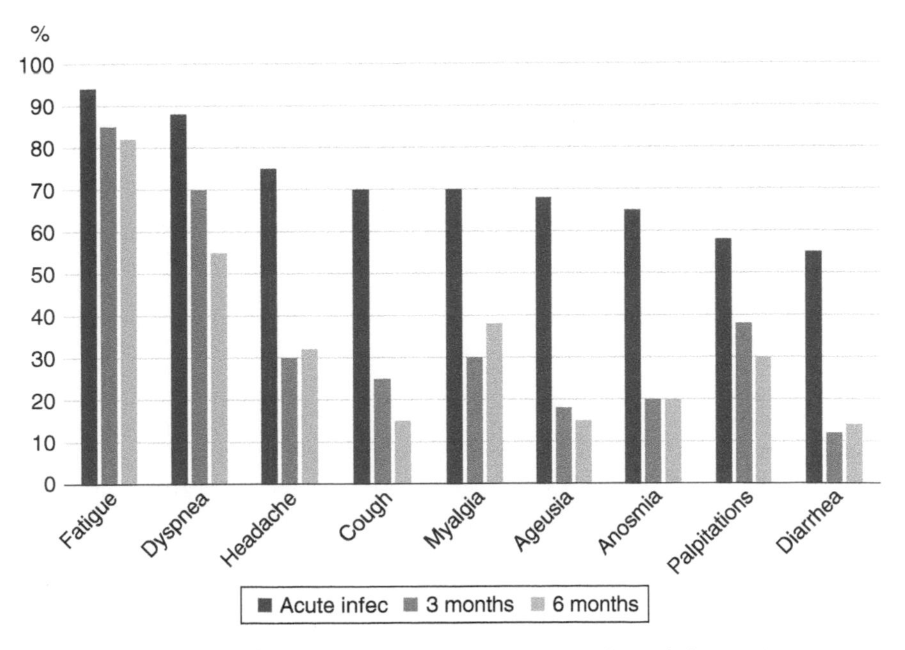

Figure 1.3 Symptoms during acute infection, at three months, and six months.
Source: Modified from Vaes et al. [18].

60% during the acute infection, decreased to 20% of subjects at both three and six months. As in hospitalized patients, the number of symptoms present initially correlated with long COVID.

Comparing Hospitalized and Non-hospitalized Patients

A systematic analysis of more than 11 000 patients with COVID-19 compared the demographic characteristics and symptoms in hospitalized versus non-hospitalized patients [20]. The demographic characteristics and long-COVID symptoms varied greatly among the two groups (Table 1.4). Non-hospitalized, long-COVID patients were younger, and three-quarters were female. The symptoms were also very different.

Neuropsychiatric symptoms, including cognitive disturbances, particularly confusion, and mood disturbances were much more common in the non-hospitalized patients. Furthermore, the prevalence of most neuropsychiatric symptoms, including depression, anxiety, and sleep disturbances increased from follow-up time points at three months to six months whereas a loss of smell, taste, myalgias, and cognitive dysfunction remained unchanged.

Table 1.4 Comparing hospitalized versus non-hospitalized COVID-19 patients.

Characteristic	Hospitalized	Non-hospitalized
Demographics		
Female (%)	46	74
Mean age (yrs)	58	45
Long-COVID symptom (%)		
Fatigue	27	50
Brain fog	23	39
Pain	27	31
Confusion	2	49
Memory issues	26	30
Myalgias	7	27
Headache	3	32
Sleep disturbance	25	36
Depression	14	25
Anxiety	17	31

Source: Based on Premraj et al. [20].

The fact that a loss of taste and smell do not change much from three to six months, compared to depression, anxiety, and sleep disturbances that all increased from three months to more than six months, suggests different illness mechanisms. This makes intuitive sense since mood and sleep disturbances are often considered psychiatric in nature, whereas a loss of taste and smell are neurologic disorders. As we will discuss throughout the book, this artificial distinction falls apart when we closely examine the combination of biological and psychological pathways involved in cognitive disturbances and fatigue.

Long COVID in Children

Initially there was little concern that long COVID was common in children [19, 21]. This was related to the fact that children are much less likely than adults to become symptomatic or require hospitalization. Subsequently, long COVID in non-hospitalized children was featured in the media. Kate Dardis, age 14, reported persistent headaches, exhaustion, tachycardia, shortness of breath, and difficulty concentrating months after a probable SARS-CoV-2 infection, “It’s really difficult. As a gymnast, I’ve always pushed through different types of pain and injury. I try to push myself, but with this whole experience, it’s just been too hard. My doctors are telling me I just have to listen to my body and take it one day at a time” [21].

However, in studies comparing children infected with SARS-CoV-2 to children not infected, long COVID has been rare. A systematic analysis on more than 20 000 children found a slight increase in symptoms consistent with long COVID compared to a control population [19]. The symptoms that were significantly more common after COVID-19 in these children included a loss of smell, headaches, and cognitive difficulties. In the largest, controlled study, 38 000 children with PCR-verified SARS-CoV-2 infections were compared to 78 000 seronegative children [22]. Compared to controls, only 1% of the infected children reported symptoms consistent with long COVID. In total, 33% of the infected children and 23% of non-infected children had more than three symptoms at three months. The most common symptoms were fatigue, loss of smell and/or taste, muscle weakness, dizziness, chest pain, and dyspnea. Concentration difficulty, headache, muscle or joint pain, and nausea were not more common in the infected children compared to controls. Cardiac and respiratory symptoms have been less common in children with long COVID than in adults. As in adults, there has been an alarming increase in persistent psychological symptoms in children and adolescents during the pandemic, but this has not differed between infected and non infected youth [23].

Long-term Outcomes

In hospitalized patients, long COVID is associated with organ damage, and correlates with increased age, ICU admission, and multiple symptoms upon admission. In non-hospitalized patients, there is no association of long-COVID symptoms with the severity of the initial infection. Being female and the number of symptoms at one month and nine months were risk factors for psychological distress [24–26]. In most studies of non-hospitalized patients, certain symptoms, such as fatigue and dyspnea, do not seem to change much over time, whereas others, notably a loss of smell and taste, fall dramatically [18].

These lingering symptoms adversely impact the quality of life in long-COVID patients, whether hospitalized or not. One-quarter of patients with long COVID had not recovered completely six to eight months after their infection, and one-half of patients reported working reduced hours seven months after their initial infection, and 20% were unable to work [27, 28]. Long-COVID patients, compared to controls, had poorer function, decreased physical activity, increased mood disturbances, and decreased vigor. Long-term outcomes are especially problematic in hospitalized patients with severe initial infections. One year after ICU hospitalization for COVID-19, more than one-half have had problems returning to work [27, 28].

In a study that carefully evaluated patients at four and eight months after SARS-CoV-2 infection, the persistent symptoms were quantified [29]. This study included 10 COVID-19 patients who were initially asymptomatic, 125 who were symptomatic but not hospitalized, and 44 who were hospitalized. Fever, chills, and upper respiratory symptoms were only present during the initial infection. Most of the symptoms characteristic of long COVID, including fatigue, shortness of breath, and cognitive disturbances, were prominent at each follow-up visit, but some peaked at 28–36 weeks. Fatigue was present in 45% at 28–36 weeks compared to 28% at 12–20 weeks, concentration problems in 40% versus 30%, and sleep problems in 30% versus 20%. More than 50% of the patients said that their symptoms at 12–20 weeks and at 28–36 weeks bothered them "a lot." At week six, 60% of patients stated that fatigue bothered them a lot, 65% at week 16, and 43% at week 32. Mood disturbances were reported by about one-third of patients at the 28- to 36-week period but were generally considered mild. Quality of life was significantly lower at 28 to 36 weeks than before COVID-19. These investigators suggested that patients with long COVID could be segregated into two subsets, those with multiple persistent symptoms and those with just a few.

In a nine-month report that included an equal number of hospitalized and non-hospitalized patients, more than one long-COVID symptom was present in 31% of patients at three months and 20% at nine months [30]. The most frequent symptoms at nine months were fatigue in 11%, dyspnea in 8%, myalgias in 7%, and a loss of taste in 4%. Risk factors for persistent symptoms at nine months included being more than 50-years old, ICU admission, and more than four initial symptoms. Fatigue resolved more slowly in hospitalized patients. At nine months, 19% of patients had impaired mental health, and being female was an independent risk factor for mental health symptoms.

At one year following hospitalization or emergency department visits, 57% of patients had long COVID, including fatigue in 19%, dyspnea in 19%, cognitive disturbances in 17%, myalgias in 11%, and sleep disturbances in 11% [1]. In a systematic review of hospitalized and non-hospitalized patients, long-COVID triggers for relapses included physical activity in 71%, stress in 60%, exercise in 54%, and mental activity in 46% [31]. Two-thirds of patients had moderate to extreme problems with daily activities at six months. Long-COVID sleep problems, fatigue, cognitive disturbances, depression, anxiety, and post-traumatic stress disorder (PTSD) were common, tracked together, and did not change significantly over the first six months following long COVID, either in hospitalized or non-hospitalized patients [31].

In looking at how disturbing each symptom was in Long-COVID patients, including in both hospitalized and non-hospitalized subjects, fatigue and cognitive disturbances were identified as very or moderately disturbing in 90% of patients (Figure 1.4) [32].

From the public's perspective, long COVID has become a major concern. In the UK, the Office for National Statistics (ONS) has surveyed the general population about COVID-19 and long COVID almost monthly. Self-reported long COVID, described as symptoms lasting for more than one month, were present in almost 2% of the UK population [33]. It was estimated that 1.3 million people in the UK were experiencing long COVID as of January 2, 2022. Of those who said they had long COVID, 60% said it had negatively affected their well-being, 40% said it had negatively affected their ability to exercise, and 30% that it had

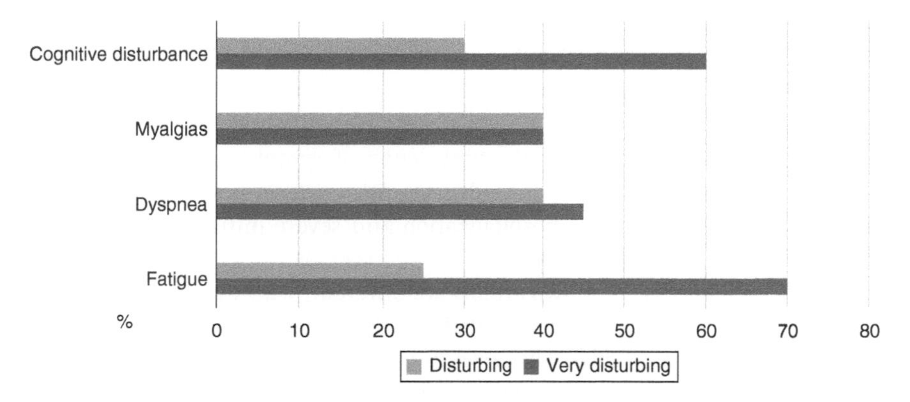

Figure 1.4 Most disturbing long-COVID symptoms. *Source:* Modified from Castanares-Zapatero et al. [32].

adversely affected their work. Compared to those who had not experienced long COVID, those with long COVID reported twofold greater depression and one-third more anxiety as well as a lower life satisfaction and happiness. One-third of patients with long COVID reported a significant adverse financial impact due to a loss of income, medical expenses, or both. One-quarter who had long COVID were struggling to pay their housing costs compared to 18% who did not have long COVID. The likelihood to self-report long COVID was greater in middle-aged females. There was a greater use of healthcare services in patients with long COVID compared to population controls, and 50% needed increased support with daily life activities, such as cleaning, preparing meals, and transportation.

Summary

There is considerable variability in prevalence estimates of long COVID, with estimates varying from 1–2% to 60–80% of infected subjects. This variability is related to many factors, most importantly the fact that there is not a unifying definition of long COVID as well as the absence of population controls in most prevalence studies. We have focused on studies that carefully controlled for symptoms in the general population. It is especially likely that self-reports are subject to recall bias.

In controlled reports, 5% to 15% of hospitalized patients and 1% to 2% of non-hospitalized patients will have long COVID, defined as three or more characteristic symptoms that last for at least three months after the initial SARS-CoV-2 infection (Table 1.1) [6, 7, 10]. We have also focused on studies in which the initial infection was documented by hospital admission for acute COVID-19 or with a positive SARS-CoV-2 test. However, in the future, other patient demographics and methodologic factors need to be assessed, as suggested by Raman and colleagues (Figure 1.5), [34].

We believe that long COVID is best appreciated as two distinct conditions: a disease, characterized by organ damage, and a syndrome, whose pathological mechanisms are unclear (Table 1.5). However, current studies on long COVID have not distinguished between the prevalence of symptoms following organ damage from those unrelated to organ damage. We have used hospitalization as a proxy for organ damage since hospitalization and severe initial disease is more likely to correlate with organ damage.

The demographics in these two groups are different, with more male patients, greater medical co-morbidities, and a significantly higher average age in the long-COVID disease patients. Long-COVID syndrome has been reported more in young, previously healthy women. In patients with long-COVID disease, compared to long-COVID syndrome, the severity of their persistent symptoms correlates with the severity of the initial infection. There are also important differences in the most

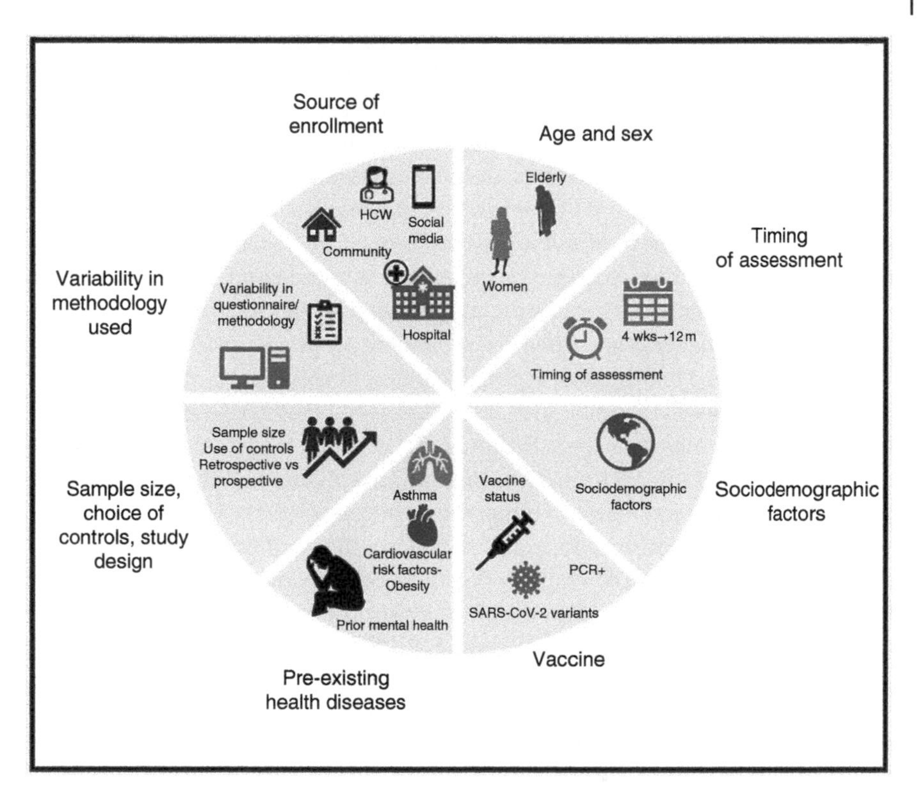

Figure 1.5 Factors that contribute to variability in prevalence estimates of long COVID. *Source:* With permission: From Raman et al. [34], Figure 2, pg 1160.

Table 1.5 Comparing long-COVID disease to long-COVID syndrome.

Characteristics	Disease	Syndrome
Average age (yrs)	50–60	30–50
Percent female	45–50	60–75
General health	Co-morbidities	Good
Hospitalized	Majority	Infrequently
Organ	Damage	Dysfunction
Correlates with initial severity	Yes	No

dominant symptoms and the symptom trajectory when comparing long-COVID disease and long-COVID syndrome at one year. Symptoms explained by organ damage, such as a loss of smell or taste, resolved in most patients between 6 and 12 months. Dyspnea has improved or stabilized, as have lung imaging and pulmonary function tests (PFTs). However, the six-minute walking time in patients with long COVID often remains prolonged in hospitalized and non-hospitalized patients [35]. This suggests that lung damage is responsible for the persistent dyspnea in most hospitalized patients but may have a more complicated etiology in non-hospitalized patients. In contrast, fatigue, widespread pain, cognitive, mood, and sleep disturbances, are often as prominent or more bothersome at 8 to 12 months than at 3 to 6 months, especially in non-hospitalized patients.

A systematic review agreed with our assessment that hospitalized and non-hospitalized patients should be differentiated, "If, in due course, significant symptomatic differences emerge from data comparing hospitalized and non-hospitalized patients, then there could be a case that the term 'long COVID' is best reserved for patients who were not hospitalized—or that a sub specifier could be useful to denote the severity of initial respiratory and/or other symptoms" [31]. Removing hospitalized patients from the long-COVID syndrome group allows us to differentiate long COVID from post-ICU syndrome (PICS), as discussed in Chapter 3.

The correlation of persistent symptoms with initial disease severity and organ damage in long-COVID disease is consistent with biomedical disease models. In keeping with usual disease trajectory, symptoms such as a loss of taste and smell or dyspnea improve over time. In contrast, the demographic differences, lack of correlation with initial severity or defined organ damage, and little improvement in symptoms such as fatigue or pain suggests a different illness model for long-COVID syndrome. In Chapters 2 and 3, we will focus on long-COVID disease and how it has primarily damaged the lungs, heart, and brain. In Sections 2 and 3 we focus on organ dysfunction and introduce biopsychological and neuro-immunologic illness models to understand long-COVID syndrome.

References

1 National Institute for Health and Care Excellence (NICE), Scottish Intercollegiate Guidelines Network (SIGN), Royal College of General Practitioners (RCGP). (2022) *COVID-19 rapid guideline: managing the long-term effects of COVID-19*.

2 Soriano, J.B., Murthy, S., Marshall, J.C. et al. (2021). A clinical case definition of post-COVID-19 condition by a Delphi consensus. *Lancet Infect. Dis.* http://doi.org/10.1016/S1473-3099(21)00703-9.

3 David, A. (2021). Long COVID: research must guide future management. *BMJ* 375: n3109.

4 Amin-Chowdhury, Z. and Ladhani, S.N. (2021). Causation or confounding: why controls are critical for characterizing long COVID. *Nat. Med.* 27: 1129.

5 Matta, J., Wiernik, E., Robineau, O. et al. (2022). Association of self-reported COVID-19 infection and SARS-CoV-2 serology test results with persistent physical symptoms among French adults during the COVID-19 pandemic. *JAMA Intern. Med.* 182: 19.

6 Ayoubkhani, D., Khunti, K., Nafilyan, V. et al. (2021). Post-COVID syndrome in individuals admitted to hospital with covid-19: retrospective cohort study. *BMJ* 372: n693. https://doi.org/10.1136/bmj.n693.

7 Huang, C., Huang, L., Wang, Y. et al. (2021). 6-month consequences of COVID-19 in patients discharged from hospital: a cohort study. *Lancet* 397: 220.

8 Chevinsky, J.R., Tao, G., Lavery, A.M. et al. (2021. 15). Late conditions diagnosed 1–4 months following an initial COVID-19 encounter: a matched cohort study using inpatient and outpatient administrative data — United States, March 1–June 30, 2020. *Clin. Infect. Dis.* 73 (Suppl 1): S5.

9 Rivera-Izquierdo, M., Láinez-Ramos-Bossini, A.J., de Alba, I.G.F. et al. (2022). Long COVID 12 months after discharge: persistent symptoms in patients hospitalised due to COVID-19 and patients hospitalised due to other causes—a multicentre cohort study. *BMC Med.* 20: 92. https://doi.org/10.1186/s12916-022-02292-6.

10 Nasserie, T., Hittle, M., and Goodman, S.N. (2021). Assessment of the frequency and variety of persistent symptoms among patients with COVID-19. *JAMA Netw. Open* 4 (5): e2111417.

11 Evans, R.A., McAuley, H., Harrison, E.M. et al. (2021). Physical, cognitive and mental health impacts of COVID-19 after hospitalization: a UK multi-Centre prospective cohort study. *Lancet Respir. Med.* 9: 1275.

12 Ayoubkhani, D., Bosworth, M., and King, S. (2021). *Prevalence of ongoing symptoms following coronavirus (COVID-19) infection in the UK: 4 November 2021*. London: Office of National Statistics.

13 Sudre, C.H., Murray, B., Steves, C.J. et al. (2021). Attributes and predictors of long COVID. *Nat. Med.* 27: 626.

14 Whitaker, M., Elliott, J., Chadeau-Hyam, M., et al. (2021). Persistent symptoms following SARS-CoV-2 infection in a random community sample of 508,707 people. http://hdl.handle.net/10044/1/89844 (accessed 20 May 2022).

15 Amin-Chowdhury, Z., Harris, R.J., Aiano, F., et al. (2022). Characterizing post-COVID syndrome more than 6 months after acute infection in adults; prospective longitudinal cohort study, England. *medRxiv*. https://doi.org/10.1101/2021.03.18.21253633

16 Havervall, S., Rosell, A., Phillipson, M. et al. (2021). Symptoms and functional impairment assessed 8 months after mild COVID-19 among health care workers. *JAMA* 325: 2015.

17 Cirulli, E.T., Schiabor Barrett, K.M., Riffle, S., et al. (2020). Long-term COVID-19 symptoms in a large, unselected population. *medRxiv*. https://doi.org/10.1101/2020.10.07.20208702

18 Vaes, A.W., Goertz, Y.M.J., Van Herck, M. et al. (2021). Recovery from COVID-19: a sprint or marathon? 6-month follow-up data from online long COVID-19 support group members. *ERJ Open Res.* 7 (2): 00141–02021.

19 Behnood, S.A., Shafran, R., Bennett, S.D. et al. (2021 Nov 20). Persistent symptoms following SARS-CoV-2 infection among children and young people: a meta-analysis of controlled and uncontrolled studies. *J. Infect.* 84 (2): P158–P170.

20 Premraj, L., Kannapadi, N.V., Briggs, J. et al. (2022). Mid and long-term neurological and neuropsychiatric manifestations of post-COVID-19 syndrome: a meta-analysis. *J. Neurol. Sci.* 434: 120162.

21 Cooney E. (2021) As more kids go down the 'deep, dark tunnel' of long Covid, doctors still can't predict who is at risk. *STAT*. (10 June).

22 Borch, L., Holm, M., Ellermann-Eriksen, S. et al. (2022 Jan 9). Long COVID symptoms and duration in SARS-CoC-2 positive children-a nationwide cohort study. *Eur. J. Pediatr.* 181: 1597–1607.

23 Stephenson, T., Pinto Pereira, S.M., Shafran, R. et al. (2022). Physical and mental health 3 months after SARS-CoV-2 infection (long COVID) among adolescents in England (CLoCK): a national matched cohort study. *Lancet Child Adolesc. Health* 6 (4): P230–P239.

24 Menges, D., Ballouz, T., Anagnostopoulos, A. et al. (2021). Burden of post-COVID-19 syndrome and implications for healthcare service planning: a population-based cohort study. *PLoS One* 16: e0254523.

25 Delbressine, J.M., Machado, F.V.C., Goertz, Y.M.J. et al. (2021). The impact of post-covid-19 syndrome on self-reported physical activity. *Int. J. Environ. Res. Public Health* 18 (11): 6017.

26 Carter, S.J., Baranauskas, M.N., Raglin, J.S., et al. (2022). Functional status, mood state, and physical activity among women with post-acute COVID-19 syndrome. *medRxiv*. https://doi.org/10.1101/2022.01.11.22269088.

27 Heesakkers, H., van der Hoeven, J.G., Corsten, S. et al. (2022). Clinical outcomes among patients with 1-year survival following intensive care unit treatment for COVID-19. *JAMA*.

28 Peluso, M.J., Kelly, J.D., Lu, S. et al. (2022). Persistence, magnitude, and patterns of postacute symptoms and quality of life following onset of SARS-CoV-2 infection: cohort description and approaches for measurement. *Open Forum Infect. Dis.* 9 (2): ofab640. https://doi.org/10.1093/ofid/ofab640.

29 Righi, E., Mirandola, M., Mazzaferri, F. et al. (2022). Determinants of persistence of symptoms and impact on physical and mental wellbeing in long COVID: a prospective cohort study. *J. Infect.* 84 (4): 566–572.

30 Maestre-Muniz, M.M., Arias, A., Mata-Vazquez, E. et al. (2021). Long-term outcomes of patients with coronavirus disease 2019 at one year after hospital discharge. *J. Clin. Med.* 10: 13.

31 Badenoch, J.B., Rengasamy, E.R., Watson, C. et al. (2021). Persistent neuropsychiatric symptoms after COVID-19: a systematic review and meta-analysis. *Brain Comm.*.

32 Castanares-Zapatero, D., Kohn, L., Dauvrin, M., et al. (2021). Long COVID: Pathophysiology – epidemiology and patient needs. Health Services Research (HSR) Brussels: Belgian Health Care Knowledge Centre (KCE). https://database.inahta.org/article/20213 (accessed 20 May 2022).

33 Ayoubkhani, D., Pawelek, P., (2022). Office of National Statistics. *Prevalence of ongoing symptoms following coronavirus COVID-19) infection in the UK: 3 February 2022.*

34 Raman, B., Bluemke, D.A., Luscher, T.F. et al. (2022). Long COVID: post-acute sequelae of COVID-19 with a cardiovascular focus. *Eur. Heart J.* 43 (11): 1157–1172.

35 Lam, G.Y., Befus, A.D., Damant, R.W. et al. (2021). COVID-19 hospitalization is associated with pulmonary/diffusion abnormalities but not post-acute sequelae of COVID-19 severity. *J. Intern. Med.* 291 (5): 694–697.

2

Lung, Heart Disease, and Other Organ Damage

Lung Disease

The lungs are the primary target of SARS-CoV-2 and the organ most often damaged, sometimes irreversibly. Most critically ill COVID-19 patients develop pulmonary disease with X-ray and imaging evidence of ground glass nodules that progress to widespread consolidation. Acute lung disease with hypoxemia is the usual cause of death in acute COVID-19 patients. Severe pulmonary disease is more common in male patients and those with multiple co-morbidities. The lung disease is characterized as an atypical acute respiratory distress syndrome (ARDS), with greater vascular involvement and relatively normal lung compliance compared to usual ARDS consolidation [1].

The most prominent autopsy findings include diffuse alveolar inflammation and vascular injury related to micro-thromboses [2]. Inflammation is manifested by a diffuse acute and chronic alveolar inflammation and sometimes a broncho-pneumonia from secondary infection [3]. Microthrombi and large thrombi with hyaline membrane changes characterize the vascular changes. Traces of SARS-CoV-2 genetic footprints in lung tissue were found but that does not correlate with progressive tissue damage.

Most reports on lung disease after SARS-CoV-2 infection focus on hospitalized patients. At the time of hospital discharge, 90% of critically ill COVID-19 patients have ongoing lung dysfunction [4]. In hospitalized patients, 25–50% still have evidence for persistent pulmonary disease at three to four months, and this organ damage is sixfold higher than that of matched controls [5].

However, at one year, less than 10% have evidence of persistent lung damage and only 5% report persistent dyspnea [6]. More detailed pulmonary function tests (PFTs) on a sub-group of these patients revealed a correlation of abnormal lung function with initial respiratory status. Adequate recovery of lung function,

Unravelling Long COVID, First Edition. Don Goldenberg and Marc Dichter.

Table 2.1 Lung function at 6 and 12 months according to respiratory difficulty during initial hospitalization.

	Not requiring oxygen		Needed ventilatory aid	
PFT	**6 Months (%)**	**12 Months (%)**	**6 Months (%)**	**12 Months (%)**
FEV1 < 80%	7	4	14	6
TLC < 80%	11	5	39	29

Source: Based on Huang et al. [7].

including the forced expiratory volume in one second (FEV1) and total lung capacity (TLC) at 6 and 12 months, correlate with admission to an intensive care unit (ICU) and whether patients required mechanical ventilatory assistance (Table 2.1) [7]. A TLC of less than 80% of the predicted TLC was still present at one year in one-third of the hospitalized patients who needed ventilatory support.

The abnormalities in lung imaging studies and PFTs usually correlate with the severity of lung disease during the initial SARS-CoV-2 infection. The most consistent PFT abnormality is a reduction in the carbon monoxide diffusion capacity (*D*LCO) [8, 9]. In a six-month follow-up study of 65 patients with severe COVID-19 pneumonia, PFTs were abnormal in more than one-half of patients, and about one-third had minimal interstitial changes on lung computed tomography (CT) scans with 5% demonstrating severe pulmonary fibrosis [10]. There is new evidence that fibrosis or fibrotic-like features on CT scans may not correlate well with pulmonary symptoms and are often reversible [11].

Objectively abnormal lung tests are often present at six months to one year after hospitalization. Longitudinal imaging was abnormal in one-third of cases eight months after hospital discharge following a severe SARS-CoV-2 infection [12, 13]. At one year, 25% of these patients had evidence of ground glass opacities and subpleural lesions on chest imaging. In most hospitalized patients, PFTs have returned to normal at one year but up to one-third continue to have exertional dyspnea [14].

Persistent small airways disease may not show up in routine PFTs and chest CT scans. These objective abnormalities can be demonstrated with inspiratory and expiratory CT scans, which have remained abnormal for 6–12 months after hospitalization [15].

Similar persistent lung disease was noted after past coronavirus infections. At 6 to 12 months following hospitalization for severe acute respiratory syndrome (SARS, caused by SARS-CoV-1), abnormal imaging and PFTs were found in up to one-third of patients [16]. Fifteen years after SARS, 5% of patients still had abnormal chest X-rays or imaging and 40% demonstrated reduced diffusion capacity on PFTs [17].

Persistent Dyspnea, Not Always Correlated with Pulmonary Damage

Often, lung damage gradually improves after severe COVID-19, even after initial structural damage. CT scan abnormalities, including ground glass opacities, often improve over time, and are no longer considered good predictors of irreversible lung damage after SARS-CoV-2 infection [18].

In many studies, persistent dyspnea did not correlate with initial disease severity or repeated PFTs [19]. In a report of 100 patients hospitalized for SARS-CoV-2 infections, at 10 months, 50% of patients reported persistent dyspnea and fatigue [20]. In those hospitalized patients, there was no correlation with PFT abnormalities, including forced expiratory volume (FEV), forced vital capacity (FVC), TLC, residual capacity (RV), pulse oxygen saturation (SpO2) or six-minute walk distance (6MWD) with persistent shortness of breath (Figure 2.1).

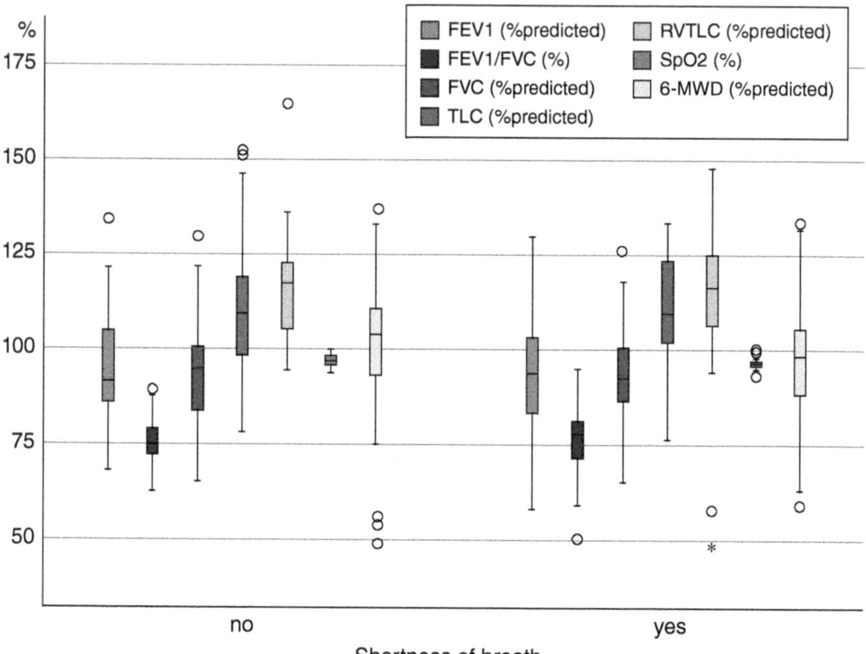

Figure 2.1 Lack of correlation between PFTs and dyspnea at 10 months post-COVID-19. *Source:* With permission. From Staudt et al. [20]. Figure 3A. Box plots of functional measures in percent predicted, or percentages for SpO_2 and FEV1/FVC, for the two groups of patients either reporting or not reporting shortness of breath at the follow-up visit. SpO_2 = oxygen saturation from pulse oximetry; FEV1 = forced expiratory volume in 1 s; FVC = forced vital capacity; TLC = total lung capacity; RV = residual capacity; 6MWD = 6-minute walk distance. The boxes indicate the quartiles, the horizontal bar the median value, the whiskers the 10- and 90-percentiles, and the circles are data points outside of these. In none of the measures were there statistically significant differences between the two groups (Mann–Whitney U-test).

An editorial in the *British Medical Journal* suggested that "while persisting lung damage may be substantial for some, for many, morbidity and mortality after COVID-19 are influenced most by pre-existing conditions, infection severity, and the extra-pulmonary complications of SARS-CoV-2" [21].

Persistent dyspnea following other coronavirus infections, such as SARS-CoV-1, also did not correlate well with objective evidence of pulmonary disease. At more than three years after hospitalization for SARS, 40% of patients had chronic fatigue and 27% met criteria for chronic fatigue syndrome (CFS) but there was no significant lung disease visible on X-ray imaging in most patients [22].

There is little data on persistent lung damage in non-hospitalized COVID-19 patients. We do know that SARS-CoV-2 positive subjects in the general population have a greater prevalence of shortness of breath six months after testing than subjects who tested negative. One report found abnormal lung imaging in 19% of hospitalized, but none of the non-hospitalized, COVID-19 patients at 75 days following infection [23]. The 6MWD was reduced in these patients but did not correlate with PFTs or imaging studies. Some authors have suggested that hospitalized patients' persistent dyspnea is related to lung damage whereas persistent symptoms such as dyspnea in non-hospitalized subjects may be related to non-pulmonary mechanisms [24].

The Heart

There is abundant evidence that viruses can attack the heart. Cardiac inflammation, called myocarditis, may follow may viral infections, including other coronaviruses as well as coxsackie virus, parvovirus, and human herpes virus 6. SARS-CoV-2 can cause an acute myocarditis with immune activation and subsequent inflammation, similar to that caused by other viruses [25]. The largest study of myocarditis found a 16-fold greater incidence of myocarditis after SARS-CoV-2 infection compared to uninfected controls [26].

SARS-CoV-2 infection in hospitalized patients is associated with several cardiac clinical manifestations that range from myocardial and pericardial inflammation, to arrythmias, heart attacks, and heart failure. Chest pain is present in 8% of patients hospitalized with acute COVID-19 and an arrythmia in 2% [27]. Fortunately, acute, new cardiac disease during SARS-CoV-2 infections is uncommon. For example, in 3000 hospitalized COVID-19 patients, only 2% had congestive heart failure, 0.5% acute coronary syndrome, and 0.1% myocarditis [28].

There are several cardiovascular sequelae in patients with long COVID. These include myocardial infarction, myocarditis, right ventricular injury, pulmonary hypertension, arrythmia, and postural orthostatic tachycardia syndrome (POTS) (Figure 2.2).

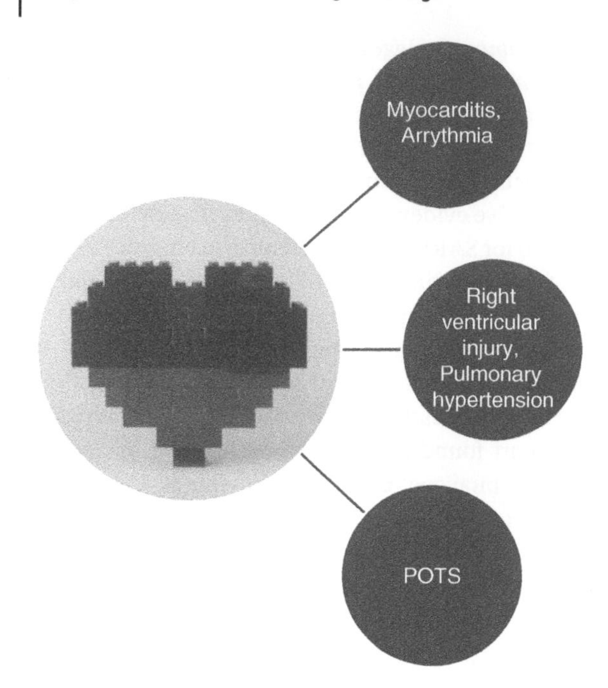

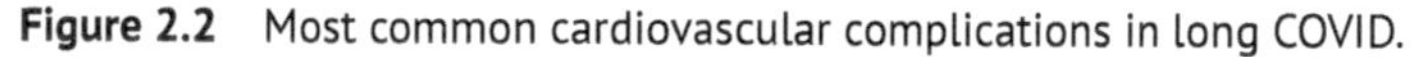
Figure 2.2 Most common cardiovascular complications in long COVID.

In a large UK study of hospitalized COVID-19 patients, at an average of 140 days post discharge, there was a threefold higher risk of heart attacks, heart failure, and arrythmia compared to matched controls who had not been infected by SARS-CoV-2 [5]. There is also an increase in stress cardiomyopathy (also known as Takotsubo syndrome), with a fourfold rise from pre-pandemic levels in one report [29]. It was postulated that the psychological, social, and economic stress associated with the COVID-19 pandemic increased the incidence of stress cardiomyopathy in hospitalized COVID-19 patients and may lead to a chronic cardiomyopathy.

An Italian study of 160 patients with SARS-CoV-2 infections, three-quarters of whom were hospitalized, found that at an average of five months most patients had persistent symptoms that correlated with functional and morphologic abnormalities [30]. These included right ventricle dilatation, increase pulmonary artery pressure, and bi-ventricular systolic-diastolic dysfunction.

In a report of 64 hospitalized and 79 non-hospitalized patients at an average of three months following infection, increased pulmonary artery pressure was found in 28% of inpatients and 20% of outpatients, abnormal diastolic function in 28% of inpatients and 20% of outpatients, and abnormal systolic function in 9% of inpatients and 8% of outpatients [31].

A systematic review of cardiac abnormalities found major cardiac events, including heart attacks, arrythmias, and heart failure at three to six months in 5% of COVID-19 patients, which was slightly increased compared to a control population [32]. Cardiac symptoms and laboratory evidence of cardiac disease was common during the first three months after acute SARS-CoV-2 infection. Chest pain occurred in 23% of patients in the first three months and 5% at three to six months; palpitations occurred in 6% of patients during the first three months compared to 9% during months three to six (Figure 2.3). An abnormal echocardiogram (ECG), cardiac biomarkers, echocardiogram, and cardiac imaging were common in the first three months. A clinical diagnosis of myocarditis was made during the first three months in 13% of patients, but only 1% had evidence of myocarditis at three to six months.

As anticipated, most patients with significant cardiac disease during acute COVID-19 had pre-existing cardiac disease. In those patients without pre-existing

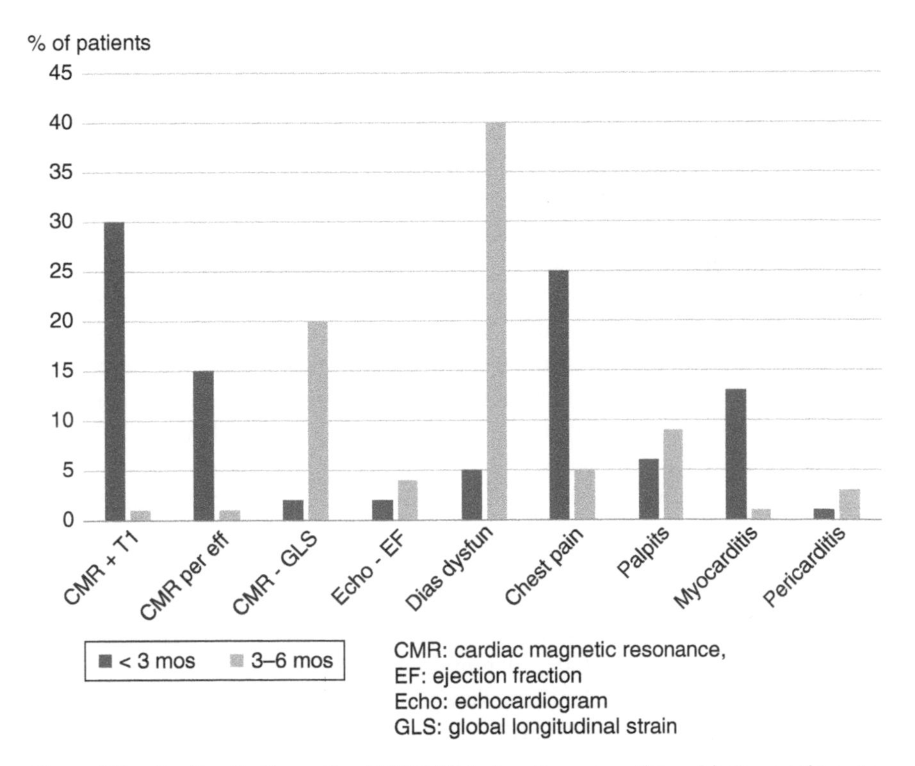

Figure 2.3 Cardiac findings after COVID-19 before three months and between three to six months. *Source:* Modified from Said Ramadan et al. [32].

cardiac disease, echocardiographic studies demonstrated an ejection fraction of less than 50% in 3%, abnormal global longitudinal strain in 24%, left ventricular diastolic dysfunction in 20%, and right ventricular diastolic dysfunction in 16% [33]. Even more subtle myocardial injury, as determined by increased high-sensitivity cardiac troponin T levels, was associated with long-COVID symptoms [34].

Is There Unanticipated Long-term Cardiac Damage?

The concern early in the pandemic was whether SARS-CoV-2 was particularly toxic to the heart, both during the acute infection and over the "long-haul" as well as whether subtle cardiac abnormalities not obvious during acute SARS-CoV-2 infection result in long-term cardiac damage. We know that physiologic cardiac abnormalities during hospitalization for acute COVID-19 are common. They most often involved the right side of the heart with right ventricular dysfunction in up to 50% of reports and are often associated with elevated pulmonary artery pressure and left ventricular systolic dysfunction [35].

There was great concern that these measures of cardiac disease during acute infection were harbingers of later-developing cardiac disease following even mild SARS-CoV-2 infections. An early autopsy study found that 24 of 39 elderly patients who died from COVID-19 lung disease had evidence of the virus present in cardiac tissue [36]. None of these patients were diagnosed antemortem with myocarditis. The average patient age was 85 years. Dirk Westermann, a cardiologist at the University Heart and Vascular Centre in Hamburg described the unusual findings, "We see signs of viral replication in those that are heavily infected. We do not know the long-term consequences of the changes in gene expression yet. I know from other diseases that it is obviously not good to have that increased level of inflammation. The question now is how long these changes persist. Are these going to become chronic effects upon the heart or are these–we hope–temporary effects on cardiac function that will gradually improve over time?" [37]. However, in a much larger autopsy study of 277 patients who died of acute COVID-19, less than 2% had evidence of myocarditis [38].

The concern that long-term cardiac abnormalities, even following mild infection, might become a major problem came to the forefront following an alarming report from Germany in July 2020. In a study of 100 COVID-19 patients at an average of 71 days after infection, 78% had cardiac abnormalities detected by imaging studies or cardiac enzymes, including elevated high-sensitivity troponin, which suggests persistent myocardial inflammation [39]. These abnormalities were found in young, healthy patients (mean age 49 years) following COVID-19, and only one-third had been hospitalized.

At the time of the study, none of these people felt ill and many had just returned from ski vacations. The lead author of the study, Dr. Valentina Puntmann, a cardiologist at University Hospital, Frankfurt, said "The fact that 78% of 'recovered' [patients] had evidence of ongoing heart involvement means that the heart is involved in a majority of patients, even if COVID-19 illness does not scream out with the classical heart symptoms, such as anginal chest pain." [39]

This report was covered by 400 international news outlets and viewed a million times [40]. The cardiac abnormalities were more prominent in these 100 patients compared to healthy, control subjects. There was no correlation between cardiac abnormalities and pre-existing conditions, cardiac symptoms at the time of initial COVID-19 diagnosis or severity of the infection.

An editorial by prominent cardiologists summarized the concern from this study, "Months after a COVID-19 diagnosis, the possibility exists of residual left ventricular dysfunction and ongoing inflammation, both of sufficient concern to represent a nidus for new-onset heart failure and other cardiovascular complications. If this high rate of risk is confirmed, . . . then the crisis of COVID-19 will not abate but will instead shift to a new de novo incidence of heart failure and other chronic cardiovascular complications. We are inclined to raise a new and very evident concern that cardiomyopathy and heart failure related to COVID-19 may potentially evolve as the natural history of this infection becomes clearer" [41].

Shortly thereafter, a study from Ohio State University found that 4 of 26 (15%) college athletes who had completely recovered clinically from SARS-CoV-2 infection had abnormalities on cardiac MRI studies [42]. This study did not compare the cardiac imaging in COVID-19 subjects to those of uninfected athletes. Furthermore, similar cardiac abnormalities were found in athletes in the past. Despite these shortcomings, this small study nearly shut down college sports. Early recommendations included extensive cardiac testing, including imaging studies, in any SARS-CoV-2 infected athlete before resuming a sport.

Dr. Haider Warraich, a cardiologist in New York City, wrote a *New York Times* opinion piece on August 17, 2020, titled "COVID-19 is creating a wave of heart disease" [43]. After discussing the German study, Warraich said, "Though the study has some flaws, and the generalizability and significance of its findings not fully known, it makes clear that in young patients who had seemingly overcome SARS-CoV-2 it's fairly common for the heart to be affected. We may be seeing only the beginning of the damage." [43]. Dr. Marc Pfeffer, a cardiologist at Brigham and Women's Hospital in Boston, also worried, "We knew that this virus, SARS-CoV-2, doesn't spare the heart. We're going to get a lot of people through the acute phase [but] I think there's going to be a long-term price to pay" [37].

In contrast to the earlier Ohio State study, University of Wisconsin investigators found myocarditis in only 1% of student athletes and none were serious [44]. A report on 800 professional athletes who had a positive SARS-CoV-2 test found that less than 1% had cardiac imaging abnormalities and none experienced cardiac issues when returning to play their sport [45].

This was followed by an extensive study on the six-month cardiac function of 19 000 athletes, including 3000 with positive SARS-CoV-2 tests [46]. Several cardiac tests were checked, including cardiac enzymes, electrocardiograms, and echocardiograms and if abnormal, cardiac imaging was done. Less than 1% of the athletes had evidence of cardiac abnormalities. In another study of 1600 U.S. athletes, 2.3% had sub-clinical myocarditis [47]. Current recommendations for athletes going back to sports after a SARS-CoV-2 infection suggest a screening protocol based on symptoms, ECG, echocardiogram, and troponin blood level with a cardiac MRI, the most sensitive test, considered only if these screening tests suggest cardiac disease [48]. Although thrombotic events are important in acute COVID-19, there is no evidence that hypercoagulability or thromboembolic disease played an important role in long COVID [49].

Currently, there is no convincing evidence that patients who were not hospitalized and had no evidence of cardiac disease during the acute phase of COVID-19 will develop new-onset cardiac damage in the months following infection. The most reassuring study was a report of 74 COVID-19 positive health care workers (HCWs) compared to 75 age and sex matched control COVID-19 negative HCWs at six months after infection [50]. No differences were found in any cardiac testing, including biomarkers, echocardiograms, and cardiac MRIs in the two groups.

A study of 102 patients, including non-hospitalized subjects, at an average of 7.2 months following SARS-CoV-2 infection compared those with persistent dyspnea, chest pain, or palpitations to those without any cardiopulmonary symptoms [51]. Hospitalization and the female sex were associated with persistent cardiopulmonary symptoms. There were no significant differences in structural or functional cardiac abnormalities, and the authors concluded that ongoing myocarditis or persistent pulmonary hypertension are not the cause of the long-standing cardiopulmonary symptoms. They found increased antibody levels and markers of inflammation, including high-sensitivity C-reactive protein (hsCRP) in patients with long-COVID cardiopulmonary symptoms, suggesting immune/inflammatory mechanisms. Nevertheless, there continues to be controversy regarding long-term cardiac dysfunction in patients who had no evidence of cardiac damage during their initial SARS-CoV-2 infection and prospective studies are sorely needed.

Other Organ Disease

Gastrointestinal (GI) signs or symptoms occurred in 10% to 20% of hospitalized COVID-19 patients [52, 53]. A higher rate of acute gastrointestinal complications, including GI bleeding and mesenteric ischemia, were found in critically ill

COVID-19 patients compared to non-COVID-19, ARDS patients [54]. Diarrhea is the most common GI symptom, present upon admission in 7%, and nausea or vomiting in 5%. Abnormal liver function studies (elevated bilirubin, AST, or ALT) are present in 14% to 20% of COVID patients.

The most common GI symptoms, including diarrhea, loss of appetite, and nausea, resolved in 90% of hospitalized COVID patients by three months but the inability to gain weight persisted in some at three to six months. In a systematic review, prolonged nausea was present in 3%, diarrhea in 4%, loss of appetite in 4%, and abdominal pain in 1.7% of patients who had COVID-19 infections [55]. However, there is no evidence that chronic liver or bowel organ damage is a consequence of SARS-CoV-2 infection.

Acute kidney disease may develop in severely ill, hospitalized COVID-19 patients. It is identical to the severe kidney disease that may accompany any other life-threatening disease. Most patients were greater than 70 years and presented with pre-existing renal disease, hypertension, diabetes, or other risk factors. This form of severe kidney disease correlates with increased COVID-19 fatality rates. Not only does diabetes and obesity increase the risk of severe COVID-19 but SARS-CoV-2 infection increases the risk of diabetes. In hospitalized patients, SARS-CoV-2 infection increases blood sugar and is associated with new-onset diabetes, often presenting as diabetic ketoacidosis that requires very large doses of insulin for control. Mechanisms postulated for this diabetogenic state include direct viral or autoimmune-induced damage of pancreatic cells as well as the more non-specific insulin resistance that may accompany any severe infection. In one study, 14% of patients with mild to moderate SARS-CoV-2 infection developed new-onset diabetes [56].

There is no evidence that COVID-19 has caused an increase in systemic rheumatic diseases, such as rheumatoid arthritis or systemic lupus erythematosus. More than 10% of long-COVID subjects report persistent musculoskeletal symptoms. These are usually listed as myalgias (muscle pain and aches) or arthralgias (joint pain). There is no evidence that such symptoms are associated with physical findings, such as joint swelling, muscle weakness, or laboratory abnormalities. It is likely that these chronic musculoskeletal symptoms, including unexplained persistent chest wall pain, are manifestations of increased generalized pain sensitivity, such as that noted in fibromyalgia.

Summary

Persistent pulmonary or cardiac symptoms in hospitalized patients correlate with organ damage. This is best characterized in the lungs since the persistent shortness of breath is often the result of severe initial lung damage. In these situations, pulmonary symptoms follow the typical trajectory for any severe lung infection. Evidence of lung damage was found in most hospitalized patients for up to three months but is present in about 5% of patients from three months to one year after discharge. Persistent lung disease, with abnormal PFTs, correlates with ICU admission, ventilatory requirements during initial hospitalization, as well as possibly elevated D-dimer, LDH, and interleukin levels. In some hospitalized and non-hospitalized patients, persistent dyspnea does not correlate with objective pulmonary pathology and there is evidence that long-term physical and mental impairment are independent from the degree of initial lung damage [57].There is an increased risk for acute pulmonary embolism in patients following SARS-CoV-2 infection, especially in long-COVID patients older than 65 years [58].

Acute cardiac disease occurs in about 5% of acute SARS-Cov-2 infections and may be associated with heart attacks, arrythmias, and heart failure. Any patient with obvious cardiac disease during hospitalization may develop persistent cardiopulmonary symptoms. Even sub-clinical cardiac disease during acute SARS-Cov-2 infection was associated with such persistent symptoms. Long COVID is associated with an increased risk of cardiovascular disease, including myocardial infarction, arrythmias, and heart failure [58]. The cause of persistent dyspnea, chest wall pain, or palpitations in patients without obvious structural cardiac abnormalities is unclear. Such unexplained, persistent cardiopulmonary, gastrointestinal, and musculoskeletal symptoms, not linked to persistent organ damage, was discussed above, "Persistent Dyspnea, Not Always Correlated with Pulmonary Damage". Chapter 3 will discuss central nervous system damage and disease.

References

1 Batah, S.S. and Fabro, A.T. (2021). Pulmonary pathology of ARDS in COVID-19: a pathological review for clinicians. *Respir. Med.* 176: 106239.

2 Borczuk, A.C., Salvatore, S.P., Seshan, S.V. et al. (2020). COVID-19 pulmonary pathology: a multi-institutional autopsy cohort from Italy and New York City. *Mod. Pathol.* 33: 2156–2168.

3 Milross, L., Majo, J., Cooper, N. et al. (2021). Post-mortem lung tissue; the fossil record of the pathophysiology and immunopathology of severe COVID-19. *The Lancet Resp Med.* 10 (1): P95–P106.

4 Gattinoni, L., Coppola, S., Cressoni, M. et al. (2020). Covid-19 does not lead to a "typical" acute respiratory distress syndrome. *Am. J. Respir. Crit. Care Med.* 201: 1299.

5 Ayoubkhani, D., Khunti, K., Nafilyan, V. et al. (2021). Epidemiology of post-COVID syndrome following hospitalisation with coronavirus: a retrospective cohort study. *BMJ* 372: n693.

6 Zhang, X., Wang, F., Shen, Y. et al. (2021). Symptoms and health outcomes among survivors of COVID-19 infection 1 year after discharge from hospitals in Wuhan, China. *JAMA Netw. Open* 4 (9): e2127403.

7 Huang, C., Huang, L., Wang, Y. et al. (2021). 6-month consequences of COVID-19 in patients discharged from hospital: a cohort study. *Lancet* 397: 220–232.

8 Bellan, M., Soddu, D., Balbo, P.E. et al. (2021). Respiratory and psychophysical sequelae among patients with covid-19 four months after hospital discharge. *JAMA Netw. Open* 4: e2036142.

9 Fortini, A., Torrigiani, A., Sbaragli, S. et al. (2021). COVID-19: persistence of symptoms and lung alterations after 3-6 months from hospital discharge. *Infection* 49: 1007–1015.

10 Bardacki, M.I., Ozturk, E.N., and Ozkarafakili, M.A. (2021). Evaluation of long-term radiological findings, pulmonary functions, and health-related quality of life in survivors of severe COVID-19. *J. Med. Virol.* 93 (9): 5574–5581.

11 González, J., Benítez, I.D., Carmona, P. et al. (2021). Pulmonary function and radiologic features in survivors of critical COVID-19. *Chest* 160 (1): P187–P198.

12 Zhang, S., Bai, W., Yue, J. et al. (2021). Eight months follow-up study on pulmonary function, lung radiographic, and related physiological characteristics in COVID-19 survivors. *Sci. Rep.* 11: 13854.

13 Pan, F., Yang, L., Liang, B. et al. (2021). Chest CT patterns from diagnosis to 1 year of follow-up in COVID-19. *Radiology* 302 (3): 709–719.

14 Fortini, A., Rosso, A., Cecchini, P. et al. (2022). One-year evolution of DLCO changes and respiratory symptoms in patients with post COVID-19 respiratory syndrome. *Infection* https://doi.org/10.1007/s15010-022-01755-5.

15 Garg, A., Nagpal, P., Goyal, S., et al. (2021). Small airway disease as long-term sequela of COVID-19: Use of expiratory CT despite improvement in pulmonary function test. *medRxiv*. https://doi.org/10.1101/2021.10.19.21265028.

16 Hui, D.S., Joynt, G.M., and Wong, K.T. (2005). Impact of severe acute respiratory syndrome (SARS) on pulmonary function, functional capacity and quality of life in a cohort of survivors. *Thorax* 60 (5): 401–409.

17 Zhang, P., Li, J., Liu, N. et al. (2020). Long-term bone and lung consequences associated with hospital-acquired severe acute respiratory syndrome: a 15-year follow-up from a prospective cohort study. *Bone Res.* https://doi.org/10.1038/s41413-020-0084-5.

18 Vijayakumar, B., Tonkin, J., Devaraj, A. et al. (2021). CT lung abnormalities after COVID-19 at 3 months and 1 year after hospital discharge. *Radiology* 303 (2): 444–454.

19 Chun, H.J., Coutavas, E., and Pine, A.B. (2021). Immuno-fibrotic drivers of impaired lung function in post-acute sequelae of SARS-CoV-2. *JCI Insight.* 6 (14): e148476.

20 Staudt, A., Jorres, R.A., Hinterberger, T. et al. (2021). Associations of post-acute COVID syndrome with physiological and clinical measures 10 months after hospitalization in patients of the first wave. *Eur. J. Intern. Med.* 95: S0953:P50–60.

21 Fraser, E. (2021). Persistent pulmonary disease after acute covid-19. *BMJ* 373: n1565.

22 Ho-Bun Lam, M., Wig, Y.-K., Wai-Man, Y.M. et al. (2009). Mental morbidities and chronic fatigue in severe acute respiratory syndrome survivors. *Arch. Intern. Med.* 169 (22): 2142–2147. https://doi.org/10.1001/archinternmed.2009.384.

23 Townsend, L., Dowds, J., O'Brien, K. et al. (2021). Persistent poor health after COVID-19 is not associated with respiratory complications or initial disease severity. *Ann. Am. Thorac. Soc.* 18 (6): 997–1003.

24 Lam, G.Y., Befus, A.D., Damant, R.W. et al. (2021). COVID-19 hospitalization is associated with pulmonary/diffusion abnormalities but not post-acute sequelae of COVID-19 severity. *J. Intern. Med.* 291 (5): 694–697.

25 Weckbach, L.T., Schweizer, L., Kraechan, A. et al. (2021). Association of complement and MAPK activation with SARS-CoV-2-associated myocardial inflammation. *JAMA Cardiol.* 7 (3): 286–297.

26 Boehmer, T.K., Kompaniyets, L., Lavery, A.M. et al. (2021). Association between COVID-19 and myocarditis using hospital-based administrative data - United States, March 2020- January 2021. *MMWR Morb. Mortal. Wkly Rep.* 70 (35): 1228–1232.

27 Mirmoeeni, S., Azari Jafari, A., Hashemi, S.Z. et al. (2021). Cardiovascular manifestations in COVID-19 patients: a systematic review and meta-analysis. *J. Cardiovasc. Thoracic Res.* 13: 181–189.

28 Linschoten, M., Peters, S., van Smeden, M. et al. (2020). Cardiac complications in patients hospitalized with COVID-19. *Eur. Heart J. Acute Cardiovasc. Care* 2048872620974605. doi: https://doi.org/10.1177/2048872620974605.

29 Jabri, A., Kalra, A., Kumar, A. et al. (2020). Incidence of stress cardiomyopathy during the coronavirus disease 2019 pandemic. *JAMA Netw. Open* 3 (7): e2014780.

30 Pela, G., Goldoni, M., Cavalli, C. et al. (2021). Long-term cardiac sequelae in patients referred into a diagnostic post-COVID-19 pathway: the different impacts on the right and left ventricles. *Diagnostics* 11: 2059.

31 Giurgi-Oncu, C., Tudoran, C., Nicusor Pop, G. et al. (2021). Cardiovascular abnormalities and mental health difficulties result in a reduced quality of life in the post-acute COVID-19 syndrome. *Brain Sci.* 11: 1456.

32 Said Ramadan, M., Bertolino, L., Zampino, R. et al. (2021). Cardiac sequelae after COVID-19 recovery: a systematic review. *Clin. Microbiol. Infect.*.

33 Pournazari, P., Spangler, A.L., Ameer, F. et al. (2021). Cardiac involvement in hospitalized patients with COVID-19 and its incremental value in outcomes prediction. *Sci. Rep.* 11: 19450.

34 Weber, B., Siddiqi, H., Zhou, G. et al. (2021). Relationship between myocardial injury during index hospitalization for SARS-CoV-2 infection and longer-term outcomes. *JAHA* 11 (1): e022010.

35 Bioh, G., Botrous, C., Howard, E. et al. (2021 No 8). Prevalence of cardiac pathology and relation to mortality in a multiethnic population hospitalized with COVID-19. *Open Heart* e001833.

36 Lindner, D., Fitzek, A., Brauninger, H. et al. (2020). Association of cardiac infection with SARS-CoV-2 in confirmed COVID-19 autopsy cases. *JAMA Cardiol.* 5: 1281–1285. https://doi.org/10.1001/jamacardio.2020.3551.

37 Cooney E. (2020). Covid-19 infections leave an impact on the heart, raising concerns about lasting damage. *STAT* (27 July).

38 Halushka, M.K. and Vander, H.R. (2021). Myocarditis is rare in COVID-19 autopsies: cardiovascular findings across 277 postmortem examinations. *Cardiovasc. Pathol.* 50: 107300.

39 Puntmann, V.O., Carerj, M.L., Wieters, I. et al. (2020). Outcomes of cardiovascular magnetic resonance in patients recently recovered from coronavirus disease 2019 (COVID-19). *JAMA Cardiol.* 5 (11): 1265–1273.

40 Mandrola, J., Foy, A., Prasad, V. (2021) Setting the record straight: there is no 'Covid heart'. *STAT* 14 May.

41 Yancy, C.W. and Fonarow, G.C. (2021). Coronavirus disease 2019 (COVID-19) and the heart-is heart failure the next chapter? *JAMA Cardiol.* 5 (11): 1216–1217.

42 Rajpal, S., Tong, M.S., Borchers, J. et al. (2020). Cardiovascular magnetic resonance findings in competitive athletes recovering from COVID-19 infection. *JAMA Cardiol.* 6 (1): 116–118.

43 Warrich, H. (2020). *Covid-19 Is Creating a Wave of Heart Disease.* New York Times.

44 Starekova, J., Bluemke, D.A., Bradham, W.S. et al. (2021). Evaluation for myocarditis in competitive athletes recovering from coronavirus disease 2019 with cardiac magnetic resonance imaging. *JAMA Cardiol.* 6 (8): 945–950.

45 Martinez, M.W., Tucker, A.M., Bloom, J. et al. (2021). Prevalence of inflammatory heart disease among professional athletes with prior COVID-19 infection who received systematic return-to-play cardiac screening. *JAMA Cardiol.* 6 (7): 745–752.

46 Moulson, N., Pelek, B.J., and Drezner, J.A. (2021). SARS-CoV-2 cardiac involvement in young competitive athletes. *Circulation* 144 (4): 256–266.

47 Daniels, C.J., Rajpal, S., Greenshields, J.T. et al. (2021). Prevalence of clinical and subclinical myocarditis in competitive athletes with recent SARS-CoV-2 infection results from the big ten COVID-19 cardiac registry. *JAMA Cardiol.* 6 (9): 1078–1087.

48 Singer, M.E., Taub, I.B., and Kaelber, D.C. (2021). Risk of myocarditis from COVID-19 infection in people under age 20: a population-based analysis. *medRxiv* https://doi.org/10.1101/2021.07.23.21260998.

49 Daher, A., Balfanz, P., Cornelissen, C. et al. (2020). Follow up of patients with severe coronavirus disease 2019 (COVID-19): pulmonary and extrapulmonary disease sequelae. *Respir. Med.* 174: 106197.

50 Joy, G., Artico, J., Kurdi, H. et al. (2021). Prospective case-control study of cardiovascular abnormalities 6 months following mild COVID-19 in healthcare workers. *JACC: Cardiov Imag.* 14 (11): 2155–2166.

51 Durstenfeld, M.S., Peluso, M.J., Kelly, J.D. et al. (2021). Role of antibodies, inflammatory markers and echocardiographic findings in post-acute cardiopulmonary symptoms after SARS-CoV-2 infection. *medRxiv* https://doi.org/10.1101/2021.11.24.21266834.

52 Parasa, S., Desai, M., Chandrasekar, V.T. et al. (2020). Prevalence of gastrointestinal symptoms and fecal viral shedding in patients with coronavirus disease 2019. *JAMA Netw. Open* 3 (6): e2011335.

53 Rizvi, A., Patel, Z., Liu, Y. et al. (2021). Gastrointestinal sequelae three and six months after hospitalization for coronavirus disease 2019. *Clin. Gastroenterol. Hepatol.* 19 (11): 2438–2440.e1.

54 El Moheb, M., Naar, L., Christensen, M.A. et al. (2020). Gastrointestinal complications in critically ill patients with and without COVID-19. *JAMA* https://doi.org/10.1016/j.cgh.2021.06.046.

55 Yusuf, F. et al. (2021). Global prevalence of prolonged gastrointestinal symptoms in COVID-19 survivors and potential pathogenesis: a systematic review and meta-analysis. *F1000Res* 10: 301.

56 Sathish, T. and Chandrika, A.M. (2021). Newly diagnosed diabetes in patients with mild to moderate COVID-19. *Diabetes Metab. Syndr.* 15: 569.

57 Evans, R.A., McAuley, H., Harrison, E.M. et al. (2021). Physical, cognitive, and mental health impacts of COVID-19 after hospitalisation (PHOSP-COVID): a UK multicentre, prospective cohort study. *Lancet Respir. Med.* 9: 1275.

58 Bull-Otterson, L., Baca, S., Saydah, S. et al. (2022). Post–COVID conditions among adult COVID-19 survivors aged 18–64 and ≥65 years — United States, March 2020–November 2021. *MMWR Morb Mortal Wkly Rep.* 71 (21): 713–717.

3

COVID-19 Direct Effects on the Central Nervous System

Introduction

One could make the case that the brain is the key to unravelling many of the mysteries of long COVID, both what we label as long-COVID disease and long-COVID syndrome. Looking at broad illness mechanisms, both (i) the biomedical model (**long-COVID disease**, defined as late symptoms from direct damage that occurred during the COVID-19 illness) and (ii) the biopsychological model (**long-COVID syndrome**, the extended symptoms not clearly caused by observed damage to the brain during the COVID-19 illness and the origin of which remains unknown), fit perfectly when examining the central nervous system's (CNSs) role in long COVID.

This chapter will focus on the direct and observable effects of COVID-19 on CNS structures and functions. This includes a review of the persistent symptoms caused by CNS damage during the acute infection by the viral infection, the immune/inflammatory response, and brain hypoxia. In some cases, the neurological symptoms during COVID-19 were transient, resolved over days or weeks, and did not linger or recur. In other cases, the direct long-term consequences of the early CNS disease (for example, symptomatic strokes or diffuse brain hypoxia) continued well after other components of COVID-19 resolved. These persistent problems continued as they would have if they had instead developed in individuals with similar brain pathology in the absence of COVID-19; as such, these symptoms are not usually considered part of the long-COVID syndrome.

Oftentimes it may be difficult to determine whether persistent neurologic symptoms are related to initial brain damage or to subsequent alterations in brain function that lead to further downstream brain impairment. In later chapters, we will try to distinguish and explain symptoms experienced by individuals with the long-COVID syndrome from those that result from clear brain damage that

Unravelling Long COVID, First Edition. Don Goldenberg and Marc Dichter.

occurred during the COVID-19 illness. This will be particularly relevant to new symptoms that occur in the absence of evidence for new organ damage.

As we analyze the direct and easily detectable brain lesions in patients with moderate to severe COVID-19, most of whom were hospitalized and under medical observation, we can often predict the courses of their lingering symptoms based on the severity of the initial injury. In many cases, we can also predict the results of more extensive medical testing to characterize the extent of damage.

There are Multiple Ways for Infections to Disturb the Nervous System

Viruses or bacteria can directly infect cells within the brain or spinal cord (encephalitis, as in some herpes virus infections), infect the membranes surrounding the brain (meningitis), or by establishing abscesses within the brain. In general, these are easy to detect by imaging studies or the analysis of small amounts of cerebrospinal fluid (CSF) obtained from a lumbar puncture (LP). If a patient succumbs to the infection, an autopsy examination of the brain identifies the cause, which may include direct observation of the offending virus or another organism in the brain.

There are several autopsy studies of patients who died with COVID-19 illnesses that tried to determine if the brain was directly infected with SARS-CoV-2, with somewhat varied results. Some investigators have observed the virus in parts of the brain, especially in cells of the olfactory system, which is involved with smell [1], and in the brainstem. Most reports indicate direct viral infection of the sensory epithelia in the upper nasal passages, which bear odorant receptors. (Epithelia are cells that line the interior surfaces of the body, including the nasal cavity. They are not part of the nervous system, nor the actual olfactory nerves.) There are also rare reports of patients exhibiting neurologic symptoms that may have been associated with encephalitis caused by the virus, or possibly by the immune system [2].

However, there is little evidence of extensive, direct viral invasion of the CNS in most studies. SARS-CoV-2 virus proteins were observed in a small number of neurons in cranial nerves and the brainstem but the "presence of these proteins in the CNS was not associated with the severity of the neuropathological changes" [3]. The conclusion derived from 43 autopsies was, "In general, neuropathological changes in patients with COVID-19 seem to be mild, with pronounced neuroinflammatory changes in the brainstem being the most common finding. There was no evidence for CNS damage directly caused by SARS-CoV-2" [3].

When viral particles were identified, there was no correlation with brain damage, as noted in an autopsy study of 41 consecutive patients with SARS-CoV-2 infections from Columbia University Irving Medical Center/New York Presbyterian Hospital [4]. The neuropathological examination of 20–30 areas from each brain revealed hypoxic/ischemic changes in all brains, both global and focal; large and small infarcts, many of which appeared hemorrhagic; and microglial activation with microglial nodules accompanied by neuronophagia, most prominently in the brainstem; and evidence for neuroinflammatory lesions. Polymerase chain reaction (PCR) analysis revealed the presence of “low, to very low” but detectable viral RNA. However, PCR testing can detect very small fragments of virus genetic material as well as intact, potentially infectious, virus. In situ hybridization and immunohistochemisty, two other tests often used to detect evidence of infection, failed to detect viral RNA or protein in the brains. Their findings “indicate that the levels of detectable virus in COVID-19 brains are very low and do not correlate with the histopathological alterations” [4]. These investigators did find microglial activation, microglial nodules, and damaged nerve cells in the majority of autopsied brains, but these were not associated with direct viral infection of brain parenchyma. Rather, it is likely they were a result of an immune/inflammatory response, which we will discuss further in Chapter 7.

There is one major issue that arose later in the pandemic as more basic research was able to be performed on both patient material and new animal models of COVID-19. The appearance in the CSF of unique antiviral antibodies and B cells that react to SARS-CoV-2 virus, indicates that these are likely generated within the intracranial cavity rather than entering from the patient’s sera during the systemic infection. This likely indicates that a viral infection of the brain does occur, but without leaving signs of damage or inflammation [5]. This will also be elaborated on in Chapter 7.

Overview

There is strong evidence that COVID-19 causes several forms of long-term brain disease. This was documented primarily in hospitalized patients. A very large analysis of the electronic medical records of more than 81 million patients containing data from 234 379 COVID-19 patients was used to determine both new-onset and exacerbation of existing neurological and psychiatric disorders at six months after the illness [6], The risks were compared to those associated with influenza and other illnesses, and the authors concluded, “Overall, COVID-19 was associated with increased risk of both neurological and psychiatric outcomes, but the incidences of these were greater in patients who had required hospitalization, and markedly so in those who had required ICU admission or had developed encephalopathy. However, the incidence and relative risk of neurological and psychiatric diagnoses were also increased even in patients with COVID-19 who did not require hospitalization in comparison to patients hospitalized with influenza” [6].

It is likely that some of the persistent neurological problems, such as cognitive difficulties, caused directly by damage during the acute illness, were exacerbated during the long recovery periods by psychosocial factors. As we discussed for lung damage, the severity of COVID-19, the ICU experience, and the subsequent problems that survivors experienced, such as sleep disturbances, depression, anxiety, fatigue, and cognitive disturbances, all likely had a clear effect on subsequent neurological symptoms after COVID-19. In the rest of Chapter 3, we will describe some of the most significant neurological and psychiatric problems that developed in the course of COVID-19 illnesses and the sequelae of these events that became part of the long-COVID illness.

Loss of Senses of Smell and Taste

One of the more striking symptoms that occurs early in the SARS-CoV-2 infection is a loss of smell and taste. This often develops even before some of the more typical symptoms such as fever, cough, and respiratory distress. It was estimated that at least 80–90% of COVID-19 patients (before the omicron variant became dominant in the fall of 2021) had a loss or significant reduction in smell. The number of patients with a loss of smell since omicron variant is less, possibly as low as 12% according to one early Norwegian study [7]. The reason for this difference in variant phenotypes has not yet been elucidated. This estimate did not include other patients who were not tested with more quantitative smell tests. It was even suggested that a loss of smell and/or taste in pre-omicron cases was a more accurate diagnostic tool than blood tests [8]. Physicians frequently used this symptom to diagnose COVID-19.

Suzy Katz wrote an article in *The New York Times* about her sudden loss of taste and smell during a mild case of COVID-19 in early 2020, before it became widely known as a symptom of a new SARS-CoV-2 infection, "I first became aware of my anosmia, as doctors call the loss of smell, as I was cutting a clove of garlic. Despite my sensitive eyes watering from the allium I chopped, I smelled nothing. I frantically grabbed a freshly sliced lemon and sniffed with similar results. It wasn't until I accidentally left a burner on in my apartment and nearly started a fire that I finally ran to see an ear, nose and throat specialist, panic-stricken about my new disability and its long-term implications. Once he completed the exam, he said there were no indications of any physical cause for my loss of smell, and that he had no real treatment options for me. He spoke of other patients who had come in with the same issue months earlier and still complained of symptoms similar to mine. He handed me instructions for scent training without any hint of enthusiasm. The instructions said that twice a day I had to smell four essential oils: eucalyptus, rose, lemon and clove, to retrain my brain to recognize these odors. I've had no luck with this, despite my diligent adherence to the program" [9].

COVID-19 is not the only event that can cause a loss of smell and taste. Nasal obstruction during upper respiratory infections, when the nasal passages are substantially blocked, by seasonal allergies, or as a result of COVID-19, can also cause a temporary loss of smell. This is due to the nasal congestion blocking the odors from reaching the receptors and producing a temporary loss of smell. Occasionally, respiratory viruses other than SARS-CoV-2 can damage the upper respiratory epithelium that line the inside of the nose and contain the odor receptors, but this is usually much less severe than what commonly occurs in COVID-19, and it is often reversed once the original illness subsides. Traumatic brain injuries, even relatively mild, or direct nose injuries, can also produce a loss of smell, again, often temporary. In older individuals, a spontaneous, unexplained loss of smell may occur.

Smell and taste are mediated by two sense organs. Smell is mediated by odor receptors on cells in the nasal epithelium. Thin processes of the olfactory nerves penetrate this epithelium, lie close to these odor-receptor bearing cells, and are stimulated when odor receptors are activated by specific chemicals. When the nerves are activated, they send signals to the brain that identify the odors we perceive. There are hundreds of different receptors that respond to unique chemicals. Many odorants activate multiple receptors, and it was estimated that humans may be able to distinguish as many as one trillion different odors.

Taste is mediated by receptors on cells in the tongue and back of the throat that are thought to react to only five different stimuli – salt, sweet, sour, bitter, and umami (savory – salty or spicy but not sweet). Taste is mediated by a combination of activating both taste and smell receptors, which causes most individuals with a loss of smell to also complain of tasteless food, whether or not the primary taste organ is significantly affected. These individuals may still be able to identify tastes such as saltiness or sweetness, but not the complex tastes of many foods.

In our discussion of neurological problems caused by COVID-19, a loss of smell and taste should be considered a nerve or organ problem since the sensory organs (odor-bearing epithelial cells) abut the endings for the olfactory nerves that transmit the odor information into the brain. The early onset of smell loss, even before other symptoms of COVID-19, the fact that odor-receptor cells have the ACE-2 receptor as well as the demonstration of virus in the epithelial, but not the nerve, cells (with possibly, rare exceptions), indicate that this symptom is directly due to a viral infection in the smell pathway to the brain. Therefore, this neurologic symptom is a direct result of COVID-19 and corresponds to classic biologic disease models. As such, once the epithelia and nerves heal, we should expect recovery.

Indeed, most studies indicate that 70–75% of patients will recover most of their sense of smell within about four weeks after the onset of COVID-19, although 25–30% continue to suffer the loss of smell and do not recover completely. Other studies suggest an even higher recovery rate with (96.1%) objectively recovered by 12 months [10]. One reason that most patients recover the sensation(s) of smell/taste is that the nasal epithelia can regenerate after injury. In addition, the olfactory neurons also regenerate, unlike the overwhelming majority of other nerve cells within the brain.

Loss of smell and taste is also an example of the brain's rewiring going astray. In this situation, the "nerve problem" may be followed by a "brain problem"! For example, objects that normally produce fragrant odors are perceived to have strange odors not usually associated with the object being smelled (parosomia). In other individuals, very unpleasant foul odors occur in the absence of any offending source (phantosmia). These distorted recoveries were reported in as many as 56% of individuals who lost their sense of smell [11].

Even when the eyes suggest that the food should smell good, in some patients, the brain sends a message that the food smells foul. This is a clear disconnect between the reality of the odor and the perception of the odor. It cannot be corrected by what you see yourself or what others tell you they smell. Those who experience phantosmia are very distressed and this abnormal condition can wreak havoc with almost every aspect of their life [12, 13]. Suzy Katz, in *The New York Times*, described her phantosmia after COVID-19, "The comforting scents of fresh ingredients simmering on the stove that had placated my anxiety during the lockdown were soon replaced by a new, perplexing symptom called phantosmia, an olfactory hallucination of smells that aren't really there. In my case, it was a phantom sense of choking gasoline or cloying baby powder that trailed me everywhere I went. It's been nine months, and I'm still unable to detect odors. And I still have bouts of those hallucinatory scents" [9]. Other COVID-19 patients described their phantom smells, "Diet drinks taste like dirt, soap and laundry detergent smell like stagnant water or ammonia, everything or virtually everything that I eat gave me a gasoline taste or smell" [14].

Loss of taste and smell during and after a SARS-CoV-2 infection can be explained by organ damage, primarily in sensory epithelia that activate nearby olfactory nerves. There also may be secondary CNS dysfunction. This involves the bidirectional impact of emotions with taste and smell. Anosmia has been linked to depression, specifically anhedonia, an inability to feel pleasure, as well as social isolation and detachment [8]. Our sense of smell and taste greatly influence our emotions. Dr. Sandeep Robert Datta, Associate Professor of Neurobiology at Harvard Medical School said, "You think of smell as an aesthetic bonus sense but when someone is denied their sense of smell, it changes the way they perceive the environment and their place in the environment. People's sense of well-being declines. It can be really jarring and disconcerting" [14]. COVID-19 patients said, "I feel alien from myself. It's also kind of a loneliness in the world. Like a part of me is missing, as I can no longer smell and experience the emotions of everyday basic living." . . ."I feel discombobulated – like I don't exist. I can't smell my house and feel at home. I can't smell fresh air or grass when I go out. I can't smell the rain" [14]. The loss of taste and smell during COVID-19 demonstrates the complicated interplay of brain damage, dysfunction, and the outside world.

Strokes

Stokes, also known as cerebrovascular accidents (CVAs), increased during the COVID-19 pandemic and are important to consider as a cause of persistent neurologic symptoms. This was primarily among hospitalized COVID-19 patients, often in the context of other severe symptoms such as respiratory failure. Strokes, if large enough, were often easily recognized and confirmed with imaging studies. However, in many cases, the patients were too ill to transport to imaging or were not moved because of potential infectivity. Thus, the full extent of the association of acute strokes with COVID-19 and their nature remains imprecise in the literature.

Strokes are acute events that cause damage to the brain by either blocking blood flow in cerebral vessels or by bleeding into or around the brain. Strokes in the general population are usually categorized by size and etiology. Large strokes are usually very symptomatic and easily recognized by physicians. They often arise suddenly, without being part of an acute disease. Some are caused by the blockage of brain blood vessels from either atherosclerotic (vessel narrowing) or embolic events, (blood clots or small particles in the blood that travel into brain blood vessels). When large vessels are affected, the results are usually easy to recognize, unless that patient is unconscious, such as when they are sedated on respirators.

Other kinds of strokes are caused by bleeding in or around the brain. Intracerebral hemorrhages are caused by bleeding into the brain from a leaky or burst vessel. A ruptured brain blood vessel can also cause bleeding around the surface of the brain, known as subarachnoid hemorrhage. Intracerebral hemorrhages are usually diagnosed by a combination of acute symptomatology and routine brain imaging, if performed. Subarachnoid hemorrhages are often diagnosed by a spinal tap that demonstrates blood in the cerebrospinal fluid.

Some strokes are relatively small (the size of a pencil eraser) because they are caused by blockages in smaller blood vessels, and these strokes may not be easily detected unless they occur in parts of the brain where a little damage produces a large deficit. These are called "lacunes" after the Greek word for a small pit or hole. They can be detected in an MRI scan. In some brain diseases, especially those with a significant inflammatory component, very small stroke-like events can occur that are harder to identify or recognize with imaging studies. They are often referred to as microvascular lesions. These types of lesions may be relatively widespread in the brain rather than a solitary event.

Strokes that increased during acute COVID-19 were primarily related to the hypercoagulable state of blood and an inflammatory effect on brain blood vessels that interfered with normal brain oxygenation. However, there were only a relatively small number of strokes in hospitalized COVID-19 patients reported. For example, a very large study of 14483 patients from 31 hospitals in four countries (USA, Spain, Egypt, and Romania) over an approximately six-month period found that only about 1.3% had strokes [15]. Ages ranged from 60 to 79, and 60% were male. Most strokes were acute ischemic strokes. In another study of consecutive patients hospitalized over several months with COVID-19, 2.4% had ischemic strokes and 0.9% had a hemorrhagic stroke. The mean age of these patients was 64 years and only one was less than 60 years. About one-half were women [16]. Given the ages and co-morbidities, including diabetes, hypertension, and cardiac conditions, it is unclear whether the frequency of strokes in hospitalized COVID-19 patients are in excess of what might have happened to this demographic population if hospitalized with other severe infections.

In the later stages of the COVID-19 pandemic, especially with the increasingly infective delta variant in the United States, much of Europe, and elsewhere, the pattern and frequency of acute strokes changed slightly. As the COVID-19 patients became younger, there was a small increase in the prevalence of strokes in younger people, mainly between the ages of 30 to 65 years. There were some alarming reports of strokes in younger people after SARS-CoV-2 infection. Overall, the extrapolated frequency is probably 2–5% of the hospitalized patients, depending on the study, and this was higher than expected in this younger age group.

In the patients who had imaging studies, CT or MRI scans revealed multiple causes for strokes, from ischemic atherosclerotic disease to emboli in brain vessels or hemorrhagic lesions. These are all patterns that could be seen in non-COVID-19 patients. It is likely that COVID-19 facilitated these strokes in otherwise vulnerable individuals for several reasons, including the stress of the illness, a hypercoagulable state, emboli forming in the body and traveling to the brain, or vascular changes in cerebral blood vessels due to the virus or immune response. In all cases, recovery from the strokes themselves would be consistent with similar patterns of recovery in the absence of COVID-19. Recovery would be compromised by any other components of the patient's disease, such as severe respiratory and cardiac issues. It is uncommon for stroke patients (who did not have massive strokes or hemorrhages) to have a progressive decline beyond the initial clinical status unless something else is happening, such as additional small strokes, or other systemic illnesses that contribute to the disability, depression, sleep disorders, and excess stress. For many stroke patients with mild to moderate strokes, improvement over time may occur. Progressive declines due to direct organ damage are much more likely after hypoxic damage (as will be described) and would be roughly proportional to the extent of the damage. However, such lingering effects after documented strokes are not considered symptoms of the long-COVID syndrome but rather as the expected sequelae of the acute brain injury, part of long-COVID disease.

Hypoxic Brain Damage and the Post-ICU Syndrome

Hypoxic brain damage is well-documented in hospitalized patients with severe COVID-19 infection and can produce lingering neurologic symptoms. The brain damage develops from diffuse and widespread oxygen deprivation. The brain, which only accounts for about 3% to 4% of the body's mass, receives about 20% to 25% of the body's blood flow. If the brain receives too little oxygen (becoming hypoxic) from damaged lungs that poorly oxygenate the blood, CNS organ damage will quickly develop. Hypoxia is distinguished from strokes because it is not clearly caused by either brain blood vessel blockage or extensive large hemorrhages. However, in either case, the brain is injured by similar underlying processes, notably oxygen deprivation. Since the causes are usually different from those of typical strokes, hypoxic brain damage is a separate category based on a systemic problem rather than a localized problem, as is seen in most strokes.

The persistent neurologic symptoms associated with brain hypoxia are often considered part of the post-ICU syndrome (PICS). PICS is defined as "new or worsening impairment in physical (ICU-acquired neuromuscular weakness), cognitive (thinking and judgment), or mental health status that arises after critical illness and persists beyond discharge from the acute care setting" [17]. PICS can occur following other ICU stays where hypoxia was not involved, but it is clearly a larger problem when hypoxia is present. The common symptoms include generalized weakness, fatigue, decreased mobility, anxious or depressed mood, sexual dysfunction, sleep disturbances, and cognitive issues (memory disturbance/loss, slow mental processing, poor concentration and so on) [17].

Patients in the ICU with COVID-19 have very high levels of cytokines and inflammatory mediators. These are the patients most likely to experience inflammatory-related organ damage, often manifested in small brain blood vessels. Among more than 9000 critically ill patients with COVID-19 admitted to ICUs, the mortality rate at three months was 31% [18]. Mortality correlated with the severity of respiratory distress as well as with older age and co-morbidities including obesity, diabetes, and cardiovascular disease.

PICS is associated with long-term physical and psychological symptoms and causes a decreased quality of life, often lasting months or even years. In one comprehensive study, five years after leaving the ICU, patients had persistent difficulty walking, with a limited 6-minute walking distance (6MWD), despite normal PFTs [19]. The long-term physical impairment of PICS includes muscle weakness, atrophy, deconditioning, and mood and cognitive disturbances. PICS survivors and their caregivers also have long-lasting socioeconomic burdens for up to five years after hospitalization [19].

Hypoxic brain damage associated with PICS causes several acute brain-related symptoms, including confusion, alterations of consciousness, delirium, and headaches. These are not symptoms of typical focal ischemic strokes but rather a diffuse brain condition called encephalopathy. If such symptoms were displayed in the absence of an acute lung or other organ failure, they would be extensively investigated with MRIs, electroencephalograms (EEGs), and LPs to identify the cause. During the early periods of the COVID-19 outbreak, these symptoms were often considered part of the severe illness and the result of severe respiratory compromise. This constellation of severe brain disease was almost always seen in severely ill, hospitalized patients with other acute COVID-19 symptoms. Most of these patients did not receive full evaluations with extensive diagnostic tests early in their hospitalizations. It became clear that patients who exhibited delirium and altered consciousness had a significantly lower likelihood of surviving the COVID-19 illness than those who did not [20]. When these very ill patients survived, the neurological symptoms were recognized and evaluated well into the hospital stay (days or weeks after presentation and prolonged hospitalizations) at a time when most of the evidence for SARS-CoV-2 virus was gone and a clear etiology for those symptoms was not apparent. Several papers describing autopsies of patients who succumbed to the acute illness, showed predominantly hypoxic lesions in many brain regions rather than either direct viral infection of the brain or clear-cut inflammatory lesions.

One study from a consortium of French hospitals reported on 37 patients with severe COVID-19 and symptoms suggestive of hypoxic damage [21]. They described multiple lesions revealed by MRI that would account for these neurological symptoms, including lesions in the medial temporal lobes (16/37), non-confluent multifocal white matter lesions with associated hemorrhagic lesions (11/27), and extensive white matter microhemorrhagic lesions (9/37). In addition, 20/37 showed intracerebral hemorrhagic lesions with more severe symptoms. A smaller, single institution study in Boston demonstrated similar MRI findings and noted that they were similar to lesions described in the past after brain hypoxia, sepsis, or intravascular coagulation [22]. One of their patients was later autopsied, and the MRI lesions "corresponded to microvascular injury characterized by perivascular and parenchymal petechial hemorrhages and microscopic ischemic lesion" [22].

A larger, detailed clinical and pathological study also demonstrated extensive brain lesions that they described as similar to lesions common in patients with extensive hypoxia, well before the COVID-19 pandemic [23]. There was no virus noted nor major inflammatory lesions in the brain parenchyma or inflammatory vasculitis. In each of these studies, and others, the lesions were not called strokes, likely because they were not associated with extensive atherosclerosis or embolic-appearing blood clots in the involved vessels. It should be noted that the types of brain damage observed in the autopsies of COVID-19 patients are similar to those observed in patients succumbing after severe illness from varying etiologies that involved respiratory dysfunction and required ICU stays, especially after being on ventilators.

Several other studies report similar findings in patients with hypoxia and encephalopathy related to possible, progressive neurodegeneration. All these studies indicate that it is likely for patients during recovery to still exhibit the cognitive and behavioral effects of these lesions, as they did in patients with comparable hypoxic damage from other causes. One large prospective study using neurodegeneration biomarkers typically elevated in the blood of patients with Alzheimer's disease, demonstrated that patients with encephalopathy during COVID-19-related hospitalizations had greater amounts of these biomarkers than patients without encephalopathy. Furthermore, the biomarker levels for both groups were substantially higher than what is commonly seen in Alzheimer's disease [24].

Delirium, defined as an "acute mental disturbance characterized by confused thinking and disrupted attention usually accompanied by disordered speech and hallucinations" is common in the presentation of more severe forms of COVID-19. Delirium is also referred to as encephalopathy, in medical records. Delirium was primarily considered a consequence of the severe disease rather than a unique new brain problem and may not have been further analyzed in the context of pulmonary or cardiac emergencies. Extensive analyses of the brains of patients with delirium, using both MRI and postmortem examinations, report extensive evidence for hypoxic brain damage, likely secondary to the pulmonary failure.

Delirium is seen in patients with any severe chronic illness, including hypoxia, serious infections, metabolic abnormalities, or drug reactions. The behavioral symptoms of delirium may overlap with those of acute psychosis, but in many patients with acute psychosis in the context of COVID-19, the differentiation is relatively clear. Psychotic, non-delirious patients exhibit hallucinations, delusions, and disordered thinking in the *absence* of other signs of medical illness. On occasion, patients with acute psychoses presented to emergency departments without any other signs of medical illness and were later found to be positive for SARS-CoV-2. Their initial medical evaluations may have been normal, especially without fever or respiratory symptoms. New-onset psychosis can also appear in symptomatic SARS-CoV-2 patients in the absence of delirium, sometimes early in the illness and sometimes later. Most of these are case reports of a few patients at a given site.

True delirium is associated with a more severe outcome of the COVID-19 disease as well as with cognitive and behavioral symptoms later during recovery and, sometimes, for prolonged periods after the acute illness. Given the hypoxic lesions noted during the disease, the later cognitive and behavioral symptoms are attributed to brain injury during the disease.

Guillian-Barre Syndrome and Other Neuropathies – Effects of COVID-19 on Peripheral Nerves

There are reports suggesting that Guillian-Barre Syndrome (GBS) may be triggered by SARS-CoV-2 infection. If this association holds up after scrutiny, it is an example of COVID-19 causing CNS damage through immune mechanisms, as discussed in Chapter 7. GBS is a syndrome caused by dysfunction in peripheral motor and sensory nerves due to a loss of the myelin sheaths that surround the nerves. Patients present with several days of increasing muscle weakness and, less commonly, with sensory findings. The weakness can progress to very serious life-threatening illness if the muscles that control breathing become very weak. GBS is caused by an autoimmune reaction whereby antibodies against an infectious agent mistakenly attack the body's sheathing of peripheral nerves (myelin) or sometimes the nerves themselves.

GBS is usually treated by immunosuppressive methods including the use of intravenous immunoglobulins that can block the autoimmune reaction or plasmapheresis to remove the harmful antibodies from the blood. About 70% of patients recover completely, and most of the other 30% will slowly recover but may retain some weakness. A small number of patients may develop a more severe, diffuse weakness. Uncommonly, GBS can be fatal, especially in older individuals with other chronic illnesses.

In the absence of COVID-19, GBS is an uncommon, but not rare, disease that affects about one to two per 100 000 people per year. GBS was described as a sequel to several viral infections, often days or weeks after the infection subsides. There are multiple forms of GBS, but the most common are caused by an antibody attack on lipid ganglioside molecules in the myelin sheaths of motor and sensory nerves.

A relatively small number of GBS cases were reported in association with COVID-19. Most occurred days or a few weeks after the disease started, but there are occasional reports of GBS onset before symptomatic COVID-19 [25]. A large series from the United Kingdom demonstrated that new cases of GBS occurring after the COVID-19 pandemic began were less frequent than those reported in the two years before the COVID-19 pandemic. Thus, the frequency of GBS did not increase during the pandemic [26].

The characteristics of GBS that occurred during COVID-19 were somewhat different from other causes of GBS in several reports. One report noted five GBS patients in 1000 to 1200 hospitalized COVID-19 patients [27]. In the studied patients, antibodies to the common causative agents (e.g. gangliosides) in sporadic, non-COVID-19 cases were only occasionally found, and it is assumed that other autoantibodies or possibly circulating cytokines were responsible. The SARS-CoV-2 virus was universally absent in CSF collected from these patients.

Another example of the possible effects of COVID-19 on peripheral nerves is small fiber neuropathy. This is a disorder of small, sensory nerve fibers that is associated with several medical conditions and may be triggered by SARS-CoV-2. It causes tingling in the fingers and/or toes, a loss or decrease of some sensation, autonomic dysregulation, and chronic pain. It is common in diabetes, some autoimmune diseases, nutritional deficiencies, and as an unexplained consequence of aging. It is also commonly reported in chronic pain disorders, including fibromyalgia, and associated with autonomic dysfunction, especially the postural orthostatic tachycardia syndrome (POTS). It is commonly diagnosed with a skin biopsy from the distal extremity to demonstrate a reduction in intraepithelial nerve fiber density [28, 29].

Other Neurological Disorders Due to Acute COVID-19

The literature contains many short reports of other neurological conditions that have been seen in small numbers of COVID-19 patients. These include meningitis, presumed autoimmune encephalitis, various forms of acute demyelinating conditions, seizures, movement disorders, headaches, and so on. In some cases, the patients present with these syndromes without the usual signs of COVID-19, such as respiratory distress, GI symptoms, or fever. Emergency room physicians have often been surprised when SARS-CoV-2 tests returned positive. More often, these conditions were seen during the course of COVID-19, often during the post-viral, immune-mediated period. It is difficult to determine if these conditions are due directly to the viral infection, the cytokine storm that occurs early in the infection when viral titers are low, or as an autoimmune event.

There is little evidence that COVID-19 is associated with the onset of a new major psychiatric diseases, such as acute psychosis, bipolar depression, or sudden severe monopolar depression. There are numerous reports describing the effects of COVID-19 on people with pre-existing bipolar illness and major depressions, but little about the direct effects of the disease, even in these individuals. Mood disturbances, including depression, anxiety, and post-traumatic stress disorder, are common in long COVID, and we include a discussion of these issues in the next section of the book directly dealing with the Long COVID syndrome. The division of neurologic versus psychiatric is an artificial distinction in many situations. For example, both strokes and psychoses result in brain dysfunction, the former relatively easy to understand while the cause of the latter remains something of a mystery. Both result in major disabling health issues. Both conditions alter brain functions. Strokes are clearly associated with destructive brain lesions, while most serious behavioral disorders are not. Both a stroke and excessive stress can produce structural changes in the brain resulting in changes in behavior or cognition. As we discuss potential underlying mechanisms of Long-COVID in Chapters 6 and 7, we will focus on what may be an artificial separation of mind and body, physical or mental.

Summary

COVID-19 damages the central nervous system, predominantly in patients with moderate to severe disease who were hospitalized, and particularly those who required ventilatory assistance and were in the ICU. Both imaging and postmortem examinations show brain damage nearly identical to that of other patients with comparably serious illnesses. The damage took several forms including diffuse hypoxia, microvascular injuries, microhemorrhages, and, less frequently, locally defined acute strokes. These events were likely more frequent than is expected in the general population [23, 30]. There is minimal evidence for direct viral damage or extensive inflammatory damage, even in those who succumbed to the illness. There were scattered reports of other findings such as encephalitis or meningitis, but these were uncommon.

From the perspective of the long-COVID disease, patients with moderate to severe COVID-19 and organ damage from severe hypoxia or focal strokes are expected to have symptoms that long outlive their recoveries from the acute disease. In addition, there are data that some of the prolonged effects could be progressive, as in chronic neurodegenerative conditions [24]. These prolonged brain-based symptoms did not occur in isolation and could clearly be exacerbated by other conditions during the long recovery period that many of these patients endured [30].

In Chapter 6 we will focus on many symptoms of long-COVID that cannot be attributed to brain damage. Many of these lingering effects are considered downstream or indirect consequences. For example, depression is a common symptom after a stroke, whether the stroke was associated with COVID-19 or not. Additionally, we will postulate that central nervous system, neuroimmune mechanisms play a pivotal role in long-COVID syndrome. This will provide a unique framework to understand how many long-COVID symptoms, such as chronic fatigue, chronic pain, mood alterations, cognitive difficulties, including "brain fog", and sleep disturbances, can develop in the absence of organ damage.

References

1 Nampoothiri, S., Sauve, F., Ternier, G., et al., (2020). The hypothalamus as a hub for putative SARS-CoV-2 brain infection. *bioRxiv*. https://doi.org/10.1101/2020.06.08.139329.

2 Moriguchi, T., Harii, N., Goto, J. et al. (2020). A first case of meningitis/encephalitis associated with SARS-Coronavirus-2. *Int. J. Infect. Dis.* 94: 55–58.

3 Matschkel, J., Lutgehetmann, M., Hagel, C. et al. (2020). Neuropathology of patients with COVID-19 in Germany: a post-mortem case series. *Lancet Neurol.* 19: 919–929.

4 Thakur, K., Miller, E., Glendinning, M. et al. (2021). COVID-19 neuropathology at Columbia University Irving Medical Center/New York Presbyterian Hospital. *Brain* 144: 2696–2708.

5 Song, E., Bartley, C.M., Chow, R.D. et al. (2021). Divergent and self-reactive immune responses in the CNS of COVID-19 patients with neurological symptoms. *Cell Rep. Med.* 2.

6 Taquet, M., Geddes, J., Husain, M. et al. (2021). 6-month neurological and psychiatric outcomes in 236,379 survivors of COVID-19: a retrospective cohort study using electronic health records. *Lancet Psychiatr.* 8: 416–427.

7 Brandal, L.T., MacDonald, E., Veneti, L. et al. (2021). Outbreak caused by the SARS-CoV-2 Omicron variant in Norway. *Eurosurveillance* 26: 50.

8 Cooper, K., Bran, D., Farruggia, M. et al. (2020). COVID-19 and the Chemical Senses: Supporting Players Take Center Stage. *Neuron* 107: 219–233.

9 Katz S. (2020). Covid stole my sense of smell. The city's not the same. *The New York Times*. 15 December.

10 Renaud, M., Thibault, C., Le Normand, F. et al. (2001). Clinical Outcomes for Patients with Ansomia 1 Year After COVID-19 Diagnosis. *JAMA Net Open* 4 (6): e2115352.

11 Reden, J., Maroldt, H., Fritz, A., and Hummel, T. (2007). A study on the prognostic significance of qualitative olfactory dysfunction. *Eur. Arch. Otorhinolaryngol.* 264: 139–144.

12 Khan, A.M., Kallogjeri, D., and Piccirillo, J.F. (2022). Growing Public health Concern of Covid-19 Chronic Olfactory Dysfunction. *JAMA Otolaryngol. Head Neck Surg.* 148 (1).

13 Burges Watson, D.L., Campbell, M., Hopkins, C. et al. (2021). Altered smell and taste: Anosmia, parosmia and the impact of long Covid-19. *PLoS One* 16 (9): e0256998.

14 Roni, R, (2021). Some Covid survivors haunted by loss of smell and taste. *The New York Times*. 2 January.

15 Siegler, J.E., Cardona, P., Arenillas, J.F. et al. (2021). Cerebrovascular events and outcomes in hospitalized patients with COVID-19: The SVIN COVID-19 Multinational Registry. *Int. J. Stroke* 16: 437–447.

16 Rothstein, A., Oldridge, O., Schwennesen, H. et al. (2020). Acute Cerebrovascular Events in Hospitalized COVID-19 Patients. *Stroke* 27: 10.

17 Rawal, G., Yadav, S., and Kumar, R. (2017). Post-intensive care syndrome: An overview. *J. Transl. Internal Med.* 5: 90–92.

18 (2021). COVID-ICU Group on behalf of the REVA Network and the COVID-ICU Investigators. Clinical characteristics and day-90 outcomes of 4244 critically ill adults with COVID-19: a prospective cohort study. *Intensive Care Med.* 47: 60–73.

19 Herridge, M., Tansey, C.M., Matte, A. et al. (2011). Functional disability 5 years after acute respiratory distress. *N. Engl. J. Med.* 364: 1293.

20 Pranato, R., Huang, I., Anthonius, L. et al. Delirium and Mortality in Coronavirus Disease 2019 (COVID-19) – A systematic review and meta-analysis. *Arch. Gerontol. Geriatrics* 95: 2021104388.

21 Kremer, S., Lersey, F., de Seze, J. et al. (2020). Brain MRI Findings in Severe COVID-19: A Retrospective Observational Study. *Radiology* 95: e1868–e1882.

22 Conklin, J., Frosch, M.P., Mukerji, S.S. et al. (2021 117308). Susceptibility-weighted imaging reveals cerebral microvascular injury in severe COVID. *J. Neurol. Sci.* 421.

23 Sieracka, J., Sieracka, P., Kozera, G. et al. (2021). COVID-19–neuropathological point of view, pathologies and dilemmas after the first year of the pandemic struggle. *Folia neuropath* 59 (1): 1–16.

24 Frontera, J.A., Boutajanout, A., Masurkar, A.V. et al. (2022). Comparison of serum neurodegenerative biomarkers among hospitalized COVID-19 patients versus non-COVID subjects with normal cognition, mild cognitive impairment, or Alzheimer's dementia. *Alzheimer's Dement* 18: 899–910.

25 Caress, J.B., Castoro, R.J., Simmons, Z. et al. (2020). COVID-19-Associated Guillain-Barre Syndrome: The Early Pandemic Experience. *Muscle Nerve* 10.

26 Keddie, S., Pakpoor, J., Mousele, C. et al. (2021). Epidemiological and cohort study finds no association between COVID-19 and Guillain-Barré syndrome. *Brain* 144: 682–693.

27 Toscano, G., Palmerini, F., Ravaglia, S. et al. (2020). Guillain–Barré Syndrome Associated with SARS-CoV-2. *N. Engl. J. Med.* 382: 2574–2576.

28 Dani, M., Dirksen, A., Taraborrelli, P. et al. (2021). Autonomic dysfunction in 'long COVID': rationale, physiology and management strategies. *Clin. Med. (Lond)* 21 (1): e63–e67.

29 Hassani, M., Jouzdani, A.F., Motarjem, S. et al. (2021). How COVID-19 can cause autonomic dysfunctions and postural orthostatic syndrome? A Review of mechanisms and evidence. *Neurol. Clin. Neurosci.* 9: 434–442.

30 Frontera, J.A., Yang, D., Lewis, A. et al. (2021). A prospective study of long-term outcomes among hospitalized COVID-19 patients with and without neurological complications. *Neurology* 426: 117486.

Section 2

Long-COVID Syndrome and Unexplained Symptoms

4

Unexplained Symptoms: Medicine's Blind Spot

Overview

Thus far, we have discussed how persistent symptoms following SARS-CoV-2 infection are often related to organ damage, primarily involving the lungs, brain, and heart. We termed this long-COVID disease. This Section 2 of the book focuses on the persistent symptoms following COVID-19 that cannot be explained by typical disease mechanisms, what we term long-COVID syndrome. Even after extensive medical investigation, there is no evidence for organ damage that explains these persistent symptoms. These are the patients who are grasping for answers.

Long COVID or long-haul COVID was terminology first used by patients to describe and explore their wide range of symptoms. Long COVID was called the first illness created through patients connecting with each other [1]. These symptoms are also the phenomenon that most healthcare professionals find challenging to understand and to manage. Physicians have suggested that these symptoms are "medicine's blind spot" [2]. The invisible, subjective nature of symptoms such as chronic fatigue or chronic pain promotes doubt about their validity and impact.

The most common, long-lasting symptoms after COVID are fatigue, pain (myalgias and headaches), cognitive disturbances (often called brain fog), dyspnea (shortness of breath), mood disturbances, and sleep disturbances. As discussed in Section 1, persistent dyspnea can usually be explained by pulmonary damage that began during the initial infection. This chapter focuses on the common long-COVID symptoms that defy current explanations.

In systematic reviews of reports on long COVID, including both hospitalized and non-hospitalized patients, fatigue was present in 47%, cognitive disturbances in 40%, dyspnea in 32%, myalgias in 25%, joint pain in 20%, headache in 18%, cough in 18%, chest pain in 15%, altered smell in 14%, and altered taste in 7% (Figure 4.1) [3].

Unravelling Long COVID, First Edition. Don Goldenberg and Marc Dichter.

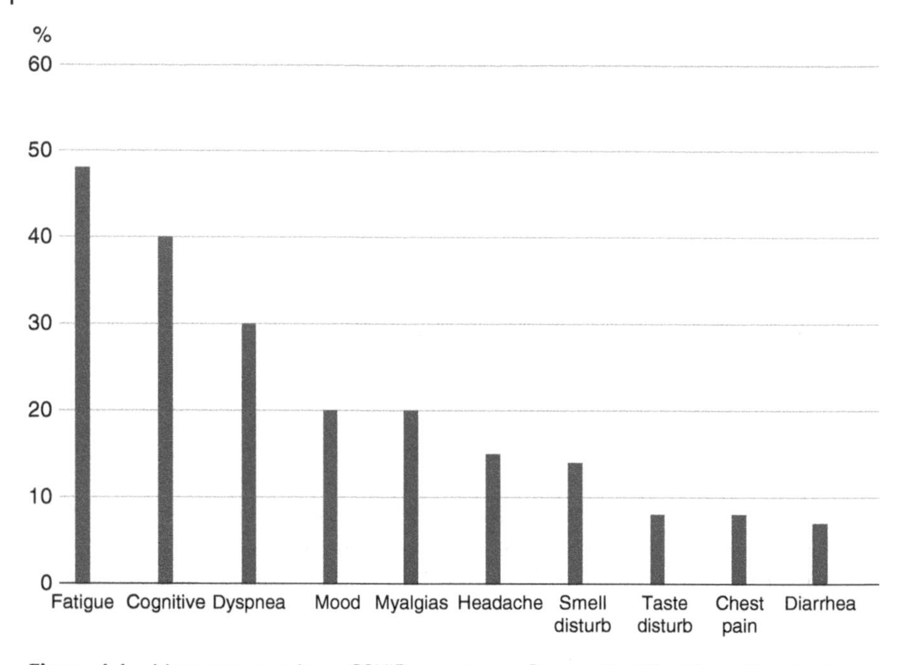

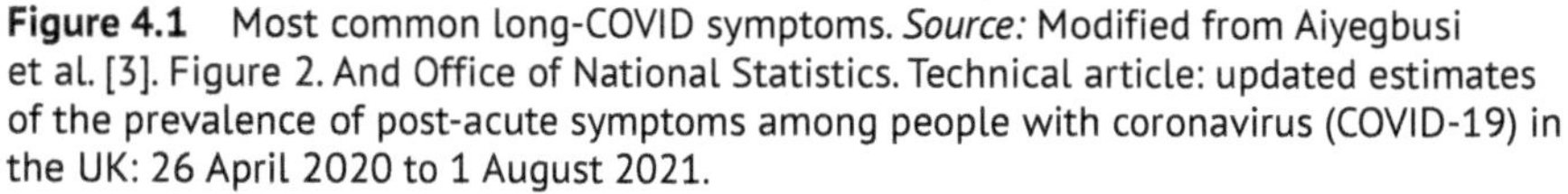

Figure 4.1 Most common long-COVID symptoms. *Source:* Modified from Aiyegbusi et al. [3]. Figure 2. And Office of National Statistics. Technical article: updated estimates of the prevalence of post-acute symptoms among people with coronavirus (COVID-19) in the UK: 26 April 2020 to 1 August 2021.

The only long-COVID symptom that correlates well with objective evidence of organ damage is dyspnea but, as discussed in Chapter 2, frequently there is little correlation between the lingering shortness of breath and abnormal chest imaging or pulmonary function tests (PFTs). We will discuss a possible explanation for that in this section.

Each of these symptoms is present at times in every human being, often most apparent during times of excess physical or emotional stress. Indeed, fatigue, back and neck pain, headaches, sleep disturbances, and mood disturbances are among the most common complaints that bring patients to health care providers. Fortunately, such common symptoms are usually transient. However, in 10% to 30% of the population these symptoms persist and typically cluster together. For example, most individuals with chronic pain also suffer from depression, the overlap is so common that it was dubbed the pain-depression dyad. People with mood disturbances usually report sleep and cognitive difficulty.

When a disease cannot be diagnosed and no organ damage is found, such chronic overlapping symptoms are often attributed to a syndrome. The label for such a syndrome is based on the most dominant symptom, such as chronic *fatigue*

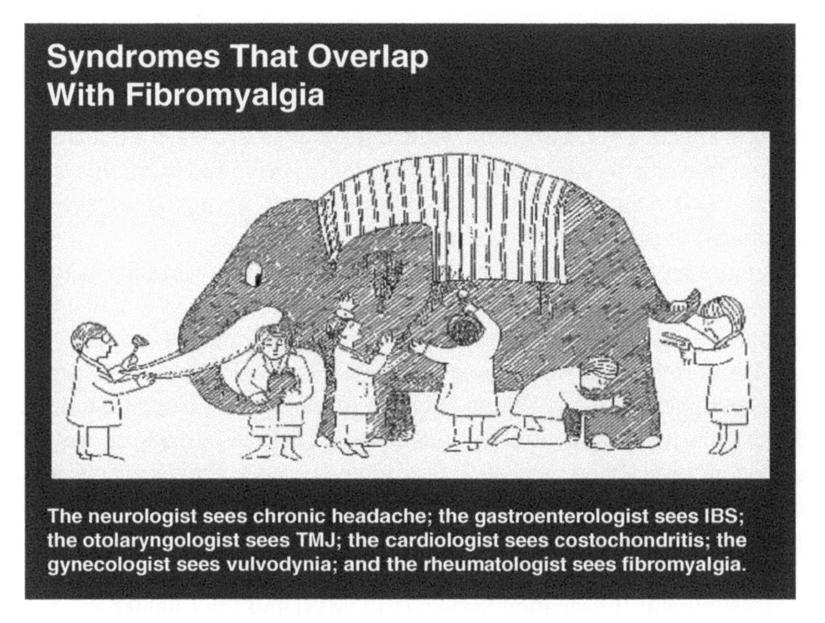

Figure 4.2 Different specialists use different syndrome names for similar symptoms. *Source:* Image from DL Goldenberg.

syndrome (CFS), fibro*myalgia* syndrome, and irritable *bowel* syndrome. Each medical specialist sees these symptoms based on their own interests. DG, a rheumatologist, will diagnose fibromyalgia; MD, a neurologist, will focus on chronic headaches; a gastroenterologist might diagnose irritable bowel syndrome; an ear, nose, and throat (ENT) specialist temporomandibular joint syndrome; a cardiologist, costochondritis; and a gynecologist, vulvodynia (Figure 4.2).

As in the general population, if there is no organ damage after SARS-CoV-2 infection, it is appropriate to refer to the persistent symptoms as part of long-COVID syndrome. Here, we will look at each of these symptoms, first in the general population, then in long-COVID syndrome. Since fatigue is the most prominent symptom in every report of long COVID, we will review its manifestations first and in some depth.

Fatigue

Chronic Fatigue in the General Population

Fatigue is one of the most common and poorly understood symptoms that plagues humans. The term is used without any agreed upon definition or accurate method of measurement. Fatigue, as most of the long-COVID symptoms, has physical and psychological components. In 1910, Sir William Osler described three elements of fatigue: (i) weakness, "the weakness as experienced by patients who have paralysis or paresis, felt in the muscles and muscle groups"; (ii) physical exhaustion, "a feeling of being knocked out, associated with slight exercise, states of exhaustion in which slight movements cause palpitation, dyspnea, perspiration, tremor and faintness"; and (iii) mental fatigue, "lack of motivation, as in many nervous patients, primarily in the cognitive domain" [4].

Physical fatigue is often thought of in terms of performing a task, such as running on a treadmill or lifting a weight. The physical output achieved is then compared to that of the average performance in an age-matched population. Cardiovascular, muscle, or nerve diseases each cause physical fatigue, sometimes referred to as peripheral fatigue. Mental fatigue is the subjective response to that task, under the control of the central nervous system. This is referred to as central fatigue, analogous to central sensitization terminology used in explaining chronic pain, as discussed below (Figure 4.3). Central fatigue is more difficult to measure, and we generally rely on validated self-report questionnaires, such as the Fatigue Severity Scale. Some of these questionnaires, such as the Patient-Reported Outcome Measurement Information System (PROMIS) and the Chalder Fatigue Questionnaire, attempt to differentiate physical, social, and cognitive fatigue but most people with chronic fatigue describe physical, emotional, and cognitive components to their exhaustion. Attempting to tease apart the physical and mental components of fatigue reflects medicine's long-running mind–body dualism. The truth is most patients suffering from chronic fatigue express both a physical and mental state of exhaustion.

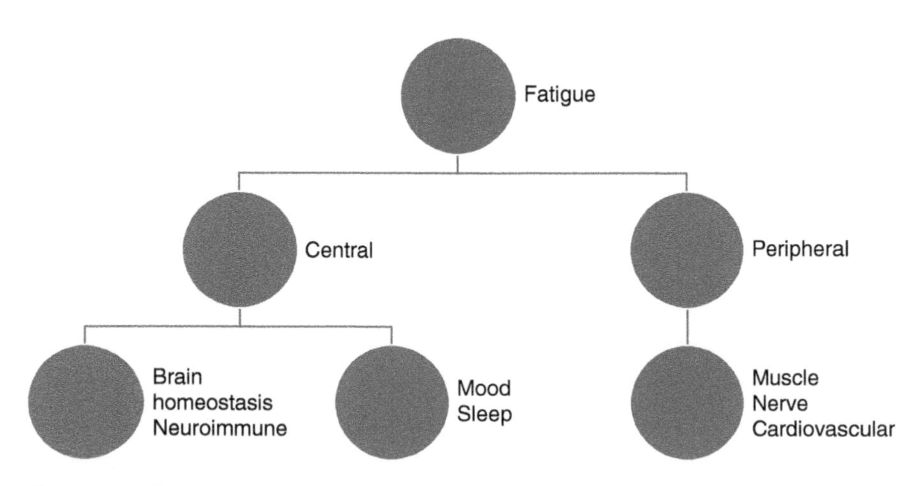

Figure 4.3 Central and peripheral components of fatigue.

Chronic fatigue is considered pathologic if it lasts longer than three months, some say longer than six months, and is severe enough to adversely affect an individual's quality of life. Medical or psychiatric diseases account for 70% to 90% of persistent fatigue in the general population [5]. Fatigue is prominent in every chronic immune and neurologic condition. For example, in multiple sclerosis (MS) fatigue is reported to be the most disturbing symptom in 50% of patients [6].

Approximately 5% to 10% of the general population has unexplained, chronic fatigue [7]. This includes patients with CFS/ME and fibromyalgia. There have been attempts to separate out those groups, such as the formal criteria for CFS/ME, which defined fatigue as lasting for more than six months and being severe enough to cause extreme difficulty getting out of bed and performing simple chores upon awakening [8].

The exhaustion that patients with illnesses such as CFS/ME and fibromyalgia describe is severe, constant, worse with exertion, and has mental and physical components. Here are some representative descriptions: "an inescapable or overwhelming feeling of profound physical tiredness, an uncontrollable, unpredictable constant state of never being rested, a ghastly sensation of being totally drained of every fiber of energy, not proportional to effort exerted, not relieved by rest, having to do things more slowly" [9].

We found this description from a young woman with CFS especially insightful, "CFS isolated me to my bed where I often had no energy to move, to talk, or even to think. If I ever had enough energy to leave my bed, it would only be for an hour or two and I would often spend double or triple that time resting up again before I could consider leaving my bed again. The fact that I spent basically all day in bed meant that by night-time, although I still had no energy to do anything, I was not physically worn out enough to sleep, meaning for some time I was practically nocturnal. Sadly, CFS is not limited to physical characteristics but can also impact people mentally and emotionally. Feeling like you are slowly losing the life as you knew it; being isolated to your room; unable to see friends/family; and missing out on activities/events can often lead to low mental health, and existing mental health issues can be exacerbated by not being able to talk to people or attend appointments. The lack of energy to take care of yourself can also harm your self-esteem" [10].

Chronic fatigue is strongly associated with mood disturbances, particularly depression, sleep disturbances, and cognitive disturbances. There is some evidence that the fatigue associated with mood and sleep disturbances improves with exercise whereas that associated with MS or CFS/ME gets worse with exercise. However, such generalizations have not been carefully validated. It is virtually impossible to separate the mental components of exhaustion from cognitive difficulty. Fatigue always has a central and peripheral component (Figure 4.3). After a stroke, more than 50% of patients report chronic fatigue and there is a strong correlation of persistent exhaustion with sleep disturbances [11]. In diseases such as cancer and MS, it is impossible to separate these components. For example, chemotherapy causes a direct physical fatigue, but subsequent exhaustion in most cancer patients has a more complicated psychological element. In MS, there is often a disconnect between brain disease on imaging studies with levels of exhaustion [12]. Brain imaging demonstrates the activation of both motor and nonmotor regions of the brain that correlate with MS patients' fatigue during either physical or mental testing [13]. We will discuss the central components of fatigue, including alterations in brain connectivity and neurohormonal changes in Section "Cognitive Disturbances".

Chronic Fatigue in Long-COVID Syndrome

Fatigue is the most prominent symptom in long COVID. If we understood chronic fatigue, we would have a good handle on long-COVID syndrome. Remember our two representative patients from Chapter 1. Both James and Sarah reported severe exhaustion long after their COVID-19 infection. James had multiple physical causes for his fatigue, beginning with his lung damage that interfered with his breathing capacity, as reflected by markedly abnormal PFTs. The months in the hospital also caused a loss of muscle mass and strength. Metabolic studies of his muscle demonstrated decreased glycogen and accumulation of lactate with evidence of muscle fiber damage. James's exhaustion can be explained by these physical factors, but he also suffered from depression as well as sleep and cognitive disturbances.

In contrast, the difficulties with daily activities described by the second patient, Sarah, are more perplexing. Her pulmonary and cardiac tests were normal, yet she had as much difficulty walking as James. Her exhaustion was also strongly associated with cognitive and mood disturbances, what she called "brain fog," the same term used by patients with CFS/ME and fibromyalgia. Since there was no physical explanation for her exhaustion, Sarah's fatigue might be classified as mental, or psychological. Yet, we have provided a model that this central fatigue is explained by alterations in brain homeostasis. Every severe illness, including long COVID, has elements of both physical (peripheral) and mental (central) fatigue, which may be related to neuroimmune dysfunction. Furthermore, these components of fatigue are not static and change over time. For example, deconditioning increases over time, augmenting the physical components of fatigue, and often precipitates greater sleep and mood disturbances, thereby increasing mental fatigue.

A systematic review of both hospitalized and non-hospitalized COVID-19 patients found that one-third of all subjects had fatigue lasting for more than three months after the initial infection [14]. This included 46% of females and 30% of males, and 34% of adults compared to 11% of children. The vast majority had persistent fatigue lasting more than six months. There was no difference in these values between hospitalized and non-hospitalized patients. Fatigue was noted in 20% to 100% of post-COVID-19 patients at three months, 16–80% at three to six months, and 30% to 80% at more than six months [14].

Some typical patient descriptions include, "Your body, it shuts down, you lift your arms up, every, everything you do, to go to the toilet, to make a cup of tea, is an effort. I haven't done anything, I haven't made a meal, haven't made a cup of tea, I tried, oh my gosh, I tried changing a pillowcase, oh my gosh, that's hard, it's so hard. No seriously, it's really hard. I folded up four jumpers. . .I folded four and I was short of breath" [15]. Another patient described her exhaustion, "it was just like I'd been run over; you know I felt, gravity felt like it was applying extra on my limbs. And I couldn't seem to manage to do anything, I stopped walking completely. And the doctor suggested, you know, post-viral fatigue and to rest, so I stopped doing anything at all" [15].

The presence of persistent fatigue does not correlate with the severity of the initial infection [16]. In a web-based survey of members in an online long-COVID peer support group, 80% reported severe fatigue at 23 weeks after a confirmed SARS-CoV-2 infection [17]. Scores were high for both physical and mental fatigue, although the physical fatigue score dropped more than the mental fatigue scores from 10 to 23 weeks. Persistent fatigue did not correlate with initial disease severity in long-term SARS survivors [18].

Cognitive Disturbances

Cognitive Disturbances in the General Population

Cognitive disturbances are common in the general population, increase with aging, and are present in many chronic medical disorders. They are almost always present in fibromyalgia and CFS, where patients commonly describe this as brain fog. They often have trouble with simple calculations, such as working with a check book or examining a restaurant bill, cannot concentrate, and have difficulty finding the right word. The symptoms vary in severity – some feel very impaired, others less so. These cognitive symptoms may be mild and annoying or serious enough for sufferers to be unable to return to work or school or even perform some routine activities. Various neuropsychological tests to assess cognitive function demonstrate attention and verbal memory deficits and impaired executive control. These are often the same tests used to quantify mental fatigue. Cognitive disturbances are strongly associated with mood disturbances, especially major depression [19]. It is difficult to determine the impact of mood and lack of focus and concentration in deciphering cognitive disturbances.

In patients with fibromyalgia and CFS/ME there is significant memory impairment, impaired executive control, and attentional dysfunction [20]. Significant cognitive impairment has been demonstrated in other chronic pain conditions including low back and neck pain and chronic headaches.

Cognitive Disturbances in Long-COVID Syndrome

Samantha Lewis described her cognitive symptoms more than one year after COVID-19 in a December 3, 2021, *New York Times* article, "I feel so stupid. I can feel that things are off. I approach a red light, my brain knows that it's red, but it's not reacting to the rest of my body to put my foot on the brake. Do you understand how terrifying that is? I was the person that fed everyone and now I struggle to figure out what I can feed myself" [21]. Her neurologist at a long-COVID rehabilitation clinic found the results of her cognitive tests to be significantly lower than average in processing speed, attention, and executive function and commented, "I was a little bit surprised with how young and functional our population was initially and how the cognitive symptoms have proved especially noticeable to people who are having demanding lives" [21].

Cognitive disturbances have been described in 15% to 40% of patients with long-COVID syndrome [22]. Another patient with long COVID, Erica Taylor, described how upset she was when she washed her TV remote with her laundry and had to return her foster dog because she could not trust herself to care for a pet, "One morning, everything in my brain was white static. I was sitting on the edge of the bed, crying, and feeling something's wrong, I should be asking for help, but I couldn't remember who or what I should be asking. I forgot who I was and where I was" [21].

Laura Holson, a *New York Times* reporter, described her own cognitive symptoms during long COVID, "More vexing was the brain fog that, for COVID survivors, can include memory loss, confusion, difficulty focusing, and dizziness. When I returned to work, I found myself losing my train of thought midsentence. On some days it felt as if words were swirling in my mind like letters in a bowl of alphabet soup being stirred with a spoon. I could see words forming, but I wasn't sure what order they should be in. One afternoon in mid-June, it took 20 minutes to write a paragraph that, on a typical day, took me a quarter of that time" [23]

Many reports did not include a control group for comparison, but a large study from Norway of non-hospitalized patients found that eight months after SARS-CoV-2 infection, 11% of infected patients, compared to 4% of uninfected subjects, reported memory difficulties [24]. In the most extensive review, cognitive impairment was present in one-quarter of patients six months after SARS-CoV-2 infection [14]. It was more prevalent in adults and in females. The prevalence of persistent fatigue did not differ between hospitalized and non-hospitalized patients.

Patients with long COVID often report that the cognitive disturbances exacerbate their physical symptoms or vice versa. One patient said, "Sometimes I feel as though if I exert myself like cognitively then my Long COVID-19 symptoms sort of exacerbate like shortness of breath, chest tightness. But like earlier on I think that it was the other way round (. . .) it seemed to be that if I exert myself physically-this means going for a five minutes walk on flat-then I get confused, I can't remember stuff, so it's like I find it really hard to unpick which way round it is" [25].

Cognitive deficits lasting from two to nine months have been commonly described in individuals with self-reported SARS-CoV-2 infections [26]. In those who had cognitive testing done, the deficits were most prominent in tests of higher order or executive functions, including reasoning, problem solving, spatial planning, and target detection. In 700 post-COVID-19 patients followed at a post-COVID-19 clinic in New York, at an average of 7.6 months after initial infection, the most prominent deficits were in processing speed (18%), executive functioning (16%), category fluency (20%), memory encoding (24%), and memory recall (23%) [27]. The relative sparing of memory recognition in the context of impaired encoding and recall suggests an executive pattern. This pattern is consistent with early reports describing a dysexecutive syndrome after COVID-19.

In a group of non-hospitalized, long-COVID patients, most cognitive abilities, including working memory, executive function, planning and mental rotation, were not different than in controls [28]. However, there were significant differences in sustained attention, as noted from tests that measured attention decrease and faster fatigue build-up over the course of a nine minute-long, attentionally demanding task. This difference could not be explained by fatigue, depression, motivation, sleep, or mood disturbances.

The most important predisposing factors for persistent fatigue and cognitive dysfunction after SARS-CoV-2 infection are pre-existing cognitive and mood disturbance [29]. At six months, non-hospitalized, long-COVID patients infrequently had impaired cognitive testing scores but there was significant correlation of cognitive performance with mood and anxiety measures [30]. The authors suggested that such findings indicate a psychological rather than a neurologic basis for the cognitive disturbances.

In contrast, it was suggested that cognitive deficits may be associated with inflammatory parameters, most often elevated cytokines rather than C-reactive protein (CRP) or D-dimer levels [14]. Most reports suggest an improvement in cognitive disturbances from months six to nine [25].

Dyspnea

In the General Population

Most often, there is a recognizable cardiac or pulmonary disease when people complain of persistent shortness of breath. This is especially the case in individuals over the age of 65 years. Unexplained dyspnea is not uncommon in younger individuals and is most often attributed to deconditioning or hyper-ventilation.

Unexplained Dyspnea in Long COVID

As reviewed in Section 1, persistent dyspnea in long COVID is often attributed to lung damage, similar to that seen in patients with acute respiratory distress syndrome (ARDS). However, often there is no evidence of cardiopulmonary disease in patients with long COVID who experience continued shortness of breath. Approximately two-thirds of long-COVID patients who experience persistent dyspnea upon exertion, palpitations, chest pain, and tachycardia will have no objective cardiac or pulmonary abnormalities with standard testing [31]. We believe that the persistent dyspnea in these patients with normal lung and heart function is likely the result of cerebrovascular, rather than cardiovascular, dysregulation. Much of this could be explained by an alteration in the central and autonomic nervous systems, which we will discuss in Chapters 5 and 6.

Deconditioning, especially after prolonged bedrest, plays a role in long-COVID persistent dyspnea. In a study of 75 hospitalized COVID-19 patients, no pulmonary mechanical limit to exercise or correlation between reduced peak exercise capacity and PFTs or chest imaging abnormalities was found [32]. The authors concluded that the reduced peak exercise capacity seen in these former COVID-19 patients was from deconditioning. Another study of 189 patients three months after hospital discharge, including 20% who were in the intensive care unit (ICU), found a reduced peak VO2 in one-third of patients, but it was not significantly different from control populations [33].

Many experts believe that the persistent dyspnea is often related to peripheral, rather than cardiopulmonary, dysfunction [34]. In one report, long-COVID patients had a reduced peak exercise aerobic capacity compared to controls that was related to impaired oxygen extraction, despite a normal cardiac output. These patients also had greater ventilatory inefficiency, creating an exaggerated hyperventilatory response during exercise (Figure 4.4) [35].

The autonomic nervous system (ANS) plays a role in long-COVID patients' dyspnea. One study found an orthostatic decline in cerebral blood-flow velocity in long-COVID patients, compared to controls, which was associated with signs of ANS dysregulation, small fiber neuropathy, and POTS [36].

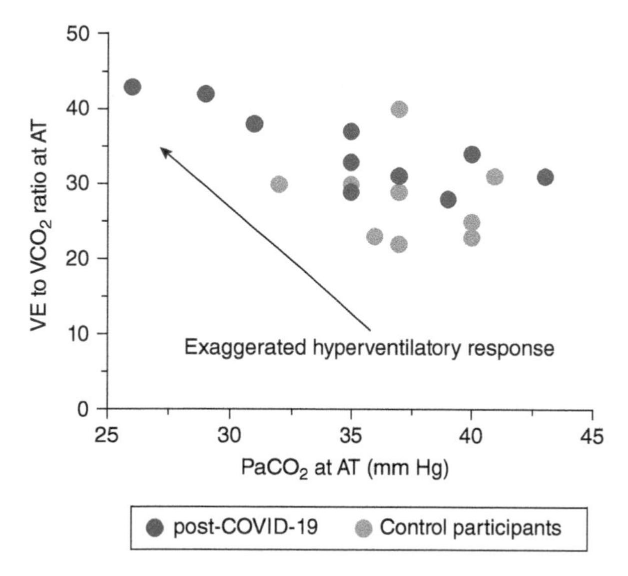

Figure 4.4 Exaggerated hyper-ventilatory response in long COVID versus controls. *Source:* With permission. Figure B1. Singh et al. [35].

Chronic Pain

Chronic Pain in the General Population

Acute pain is our body's mechanism to warn us of potential injury. It is a necessary adaptive response that protects us. Our response to acute pain is immediate and requires no conscious thought, such as withdrawing our hand when we touch something hot. In contrast, chronic pain is a pathologic process. It impairs our quality of life and can harm us. Chronic pain is the leading cause of work loss in the world. It affects 50 million people in the United States (US), 20% of the population, at a yearly cost of US$600 billion [37].

The two most common forms of chronic pain are headaches and musculoskeletal pain. A brain tumor or stroke may cause severe headaches, but most people with chronic headaches have no brain disease. Structural damage from arthritis or a muscle injury causes chronic pain, but often there is not a correlation of the joint or tissue injury with the pain severity. The brain strongly influences pain perception, as recognized in the current definition of pain, "An unpleasant sensory and emotional experience associated with actual or potential tissue damage, or described in terms of such damage . . . Pain is always subjective . . . It is unquestionably a sensation in a part or parts of the body, but it is also always unpleasant and therefore also an emotional experience" [38]

Table 4.1 Peripheral or central pain.

Feature	Peripheral	Central
Involved organ	Joint, bone, muscle, nerve	Brain
Pathology	Inflammation, damage	Dysfunction
Example	Rheumatoid arthritis, MS	Fibromyalgia

Chronic pain is often divided as to being primarily peripheral or central (Table 4.1). Peripheral sources include joints, bones, muscles, nerves, and internal organs. When there is no evidence of any peripheral, organ disease to explain the persistent pain, it is classified as central or nociplastic pain [39]. Chronic low back pain and chronic headaches are examples of central or nociplastic pain. Approximately 5% to10% of the general population have widespread muscle pain but no evidence for arthritis or any underlying musculoskeletal disease [39]. Many of these individuals are diagnosed with fibromyalgia, which is considered the prototype of central pain or central sensitization and will be discussed in Chapter 5.

Chronic pain may be considered a disease in and of itself. The burden of chronic pain includes its impact on mental health, sleep, cognition, and energy. In 2011, the Institute of Medicine surmised, "Chronic pain is the result of biological, psychological, and social factors, has distinct pathology and can be a disease in itself . . . It is important to prevent the transition from the acute to the chronic state of pain through early intervention" [39]. As we will discuss more in Chapter 5, chronic pain is linked to structural and functional changes in the brain, best exemplified by neuroimaging in patients with fibromyalgia and chronic low back pain.

Chronic Pain in Long-COVID Syndrome

A medical columnist, Kate Nicholson, described one patient's chronic pain after recovering from COVID-19, "After his second hospitalization for acute COVID-19, Tony Marks expected to get better. Then pain invaded the 54-year-old software executive's arms and legs. At first, he felt like he was covered by deep bruises, although nothing was visible on his skin. These days he told me he feels like he's being beaten repeatedly with a baseball bat. Eight months after his positive test for COVID, he said, 'Nothing touches my pain. Nothing'" [40].

If we lump muscle pain and headaches together, persistent pain is one of the most common symptoms in long-COVID syndrome. In one report, chronic pain, including myalgias and headaches, was present in one-third of patients with long-COVID syndrome six months after the initial infection [41]. The symptom of generalized pain was significantly higher in COVID-19 patients than in subjects who were followed for six months after influenza. In the long-COVID patients, muscle pain and headaches were more common in women and in younger patients.

The pain in long COVID often has a neuropathic description, including burning, numbness, and tingling that is present throughout the body. There is no clinical, laboratory, or radiologic evidence for a new form of arthritis or muscle disease in patients with long COVID. We believe that the pain reported in long COVID fits best with the that of fibromyalgia. One study found that one-third of patients with long COVID met diagnostic criteria for fibromyalgia [42]. Generalized body pain is present in 10% to 30% of patients with long-COVID syndrome [22]. Other investigators found an association between sleep disturbances and muscle pain in patients with long-COVID symptoms [43].

Headaches

Headaches in the General Population

Headaches are the most common chronic pain disorder. Tension headaches present at some point in all of us but are persistent in 5% to 20% of the general population [44]. Chronic migraine headaches occur in 5% to 10% of the population [45]. Female individuals are more prone to both types of headaches. Chronic headaches are considered disorders of the central nervous system (CNS), and there is evidence that migraine headaches are related to alterations in nerve impulses in the brain [45].

Tension and migraine headaches are associated with generalized pain, including fibromyalgia, as well as sleep disturbances and mood disturbances [46]. In one report, 70% of migraine subjects had fibromyalgia compared to 26% with tension headaches [47]. Migraine patients are two to four times more likely to have comorbid major depression than healthy controls [48]. Widespread heightened pain sensitivity is prominent in patients with migraine headaches and correlates with headache frequency and severity as well as depression.

Headaches in Long-COVID Syndrome

Headaches are one of the most common symptoms in patients with long COVID, with reports ranging from 10% to 40%. However, since headaches are so common in the general population, it is difficult to determine whether they are more prominent after COVID-19 [49]. Patients with pre-existing headaches have experienced an exacerbation of both tension and migraine headaches for months after SARS-CoV-2 infection [50]. About 20% to 30% of patients developed new headaches after COVID-19, most commonly of a severe, migraine type, bifrontal and temporal, which resolved within one month in two-thirds of patients [50].

Sleep Disturbances

Sleep Disturbances in the General Population

Twenty percent of the general population report persistent sleep disturbances. Disrupted sleep may include an inability to fall asleep, stay asleep, frequent nighttime awakenings, or difficulty awakening. There are many well-characterized sleep disorders, such as narcolepsy, sleep walking, obstructive sleep apnea, restless legs syndrome, and others. Sleep disturbances are prominent in most immune and neurological conditions.

Sleep disturbances in the general population are strongly associated with mood disturbances and chronic pain. Two-thirds of patients with chronic pain report sleep disturbances [51]. Chronic low-back pain (CLBP) subjects and those with fibromyalgia demonstrate poorer overall sleep and less efficient sleep. Patients with chronic headaches have a high prevalence of insomnia, daytime sleepiness, and snoring in comparison to healthy controls [51]. Patients with depression have a later sleep onset, longer time in bed, and a lower sleep efficiency. Sleep disturbances adversely impact immunity. For example, individuals with shorter sleep duration are at greater risk for developing a cold after viral exposure (Figure 4.5) [52].

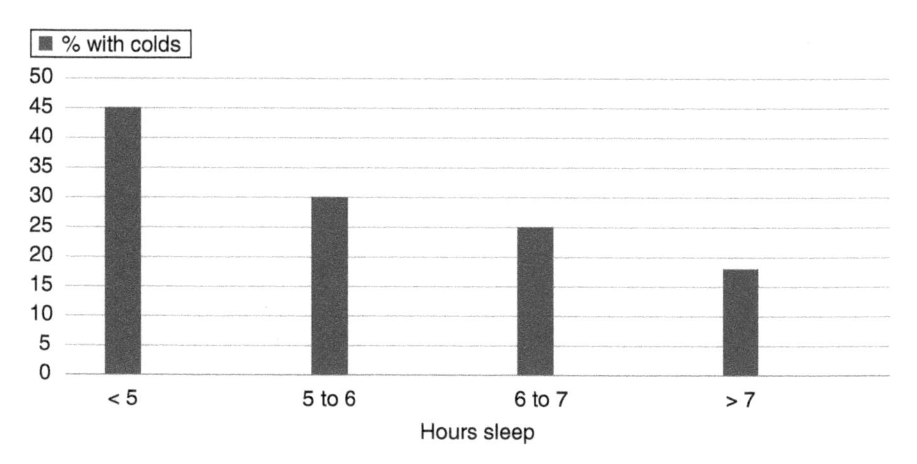

Figure 4.5 Correlation of sleep duration with developing a cold. *Source:* Modified from Prather et al. [52].

Sleep deprivation increases pain sensitivity, which was demonstrated with brain imaging that revealed changes in functional reactivity within the somatosensory cortex and thalamus [51].

Sleep Disturbances in Long-COVID Syndrome

In a systematic review, sleep disturbances were present in 20% to 30% of patients with long COVID [22]. A review of more than 2000 cases of long COVID found that on average, 47% of patients had sleep disturbances [34]. Another systematic analysis of 4318 patients with COVID-19, reported sleep disturbances in 48% and anxiety or depression in 38% [53]. As in the general population, sleep disturbances in long-COVID syndrome were associated with persistent depression and anxiety [43].

Mood Disturbances

In the General Population

Mood disturbances, mainly depression and anxiety, are present in 20% to 25% of the general population at any given point in time [54]. Major depression affects about 15% of the US population and 350 million people globally. The lifetime prevalence of depression is 8% to 10% and anxiety 12% to 15%. Post-traumatic stress disorder (PTSD) is also present in about 10% and is associated with depression and anxiety. There is a striking increase in mood disturbances for any person with chronic pain or persistent sleep disturbances [54]. Depression is also associated with cognitive disturbances, and cognitive dysfunction is a strong predictor of adverse functional outcome in individuals with depression. These overlapping features, and the increase in mood disturbances in the general population during the pandemic, makes it very difficult to determine the relationship of depression and anxiety with long COVID.

Mood Disturbances in Long COVID

In long-COVID patients, anxiety and depression were present in 20% to 50% of cases, and PTSD in 15% to 35% of cases [22]. Many of these studies lacked controls. However, depression and anxiety were twofold greater after COVID-19 compared to uninfected controls [55]. In 240 long-COVID patients, 80% female, mood disturbances were common at three and six months after a documented SARS-CoV-2 infection [56]. At three months, 47% had symptoms of depression, 37% of PTSD, and 36% of anxiety. At six months, the percentages were 41% for depression,

27% for PTSD, and 35% for anxiety. There was no significant difference in the prevalence of mood disturbances between hospitalized and non-hospitalized patients.

Following COVID-19, patients without prior mood disturbances frequently reported symptoms of depression with prominent anhedonia, a loss of interest or pleasure [57]. One report found more persistent mood and cognitive disturbances in 50 post-COVID-19 patients compared to 50 healthy controls, and none of these subjects had prior major mood or cognitive complaints [58]. The average age was 30, and 60% to 70% were female individuals, they were not hospitalized and there was no relationship between the persistent mood disturbances and the severity of infection. The depressive symptoms emerged within four months of the infection. In post-acute COVID-19 clinics, more than one-third of patients had moderate depression [59].

A group of patients with long COVID were given the Personality Assessment Inventory (PAI), an instrument that measures psychological distress [60]. The long-COVID group had high scores on somatic preoccupation and depression. Anxiety in long-COVID subjects correlated with multiple co-morbidities and was more common in female patients [61]. Increased age correlated with a reduced prevalence of anxiety. Female patients were also more likely to report sleep disturbances and decreased appetite.

Summary

In patients with long COVID, dyspnea is the only symptom that can be readily explained by well-defined organ damage. Currently, there is no evidence that other common long-COVID symptoms, including fatigue, cognitive disturbances, widespread pain, headaches, and sleep and mood disturbances, are the result of persistent infection or inflammation, or that they represent a unique, new disease state. Rather, these symptoms are akin to the same symptoms present in the general population. They almost always cluster together, and their severity is similar in individual patients [62]. The fact that these symptoms track together suggests common pathophysiologic mechanisms. These mechanisms are controlled by the CNS and involve internal and external forces. We focused on the symptom of fatigue since it so clearly demonstrates mental and physical components (Figure 4.6).

Unfortunately, medicine has been mired by what *The New York Times* medical columnist Jane Brody described as "the medical profession's artificial separation of mental and physical ills. Rather, mind and body form a two-way street. What happens inside a person's head can have damaging effects throughout the body, as well as the other way around. An untreated mental illness can significantly increase the risk of becoming physically ill, and physical disorders may result in behaviors that make mental conditions worse [54]."

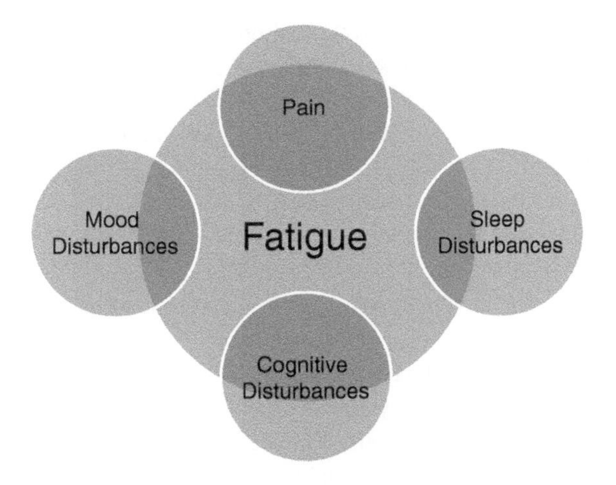

Figure 4.6 Fatigue and its intersecting symptoms in the general population, CFS/ME, and fibromyalgia, and long-COVID syndrome.

Too often, patients and their doctors conclude that any discussion about mind–body interactions brings them back to a psychosomatic or psychiatric explanation, rather than an organic disease. Drs. David, Wesseley, and Pelosi in 1988, described the dilemma of understanding psychological factors in CFS/ME, "Why have sufferers and their doctors been so vigorous in rejecting the possibility that these ubiquitous psychological factors may be aetiological in the post-viral fatigue syndrome? Firstly, patients have encountered unhelpful and even hostile responses from doctors who believe that psychiatric illnesses are not real illnesses, as is clearly portrayed in first person accounts. The humiliation perceived in the attachment of a psychiatric 'label' reflects poorly not just on these doctors but on psychiatrists, who have failed to influence such attitudes in both their colleagues and the public" [63]. In Section 3, we will explore pathways that connect these psychological and physical symptoms and how they may jointly go awry, causing the symptoms that are difficult to explain in patients with long-COVID syndrome. In Chapter 5, we will look in more depth at CFS/ME, fibromyalgia, and related disorders to provide some guidance in understanding long-COVID syndrome.

References

1 Callard, F. and Perego, E. (2021). How and why patients made Long Covid. *Soc. Sci. Med.* 268: 113426.

2 Phillips, S. and Williams, M.A. (2021). Confronting our next national health disaster-Long haul Covid. *N. Engl. J. Med.* 385: 577–579.

3 Aiyegbusi, O.L., Hughes, S.E., Turner, G. et al. (2021). Symptoms, complications and management of Long COVID: a review. *J. Royal Soc. Med.* 114 (9): 428–442.

4 Osler, W. (1910). A system of medicine. In: *Diseases of the nervous system*, vol. VIII (ed. A. Church and J.L. Salinger), 33. London: Oxford University Press.

5 Jason, L.A., Jordan, K.M., Richman, J.A. et al. (1999). A community-based study of prolonged fatigue and chronic fatigue. *J. Health Psychol.* 4: 9.

6 Bakshi, R. (2003). Fatigue associated with multiple sclerosis: diagnosis, impact and management. *Mult. Scler.* 9: 219.

7 Jason, L.A., Evans, M., Brown, M. et al. (2010). What is fatigue? Pathological and nonpathological fatigue. *PM&R* 2: 327.

8 Fukuda, K., Dobbins, J.G., Wilson, L.J. et al. (1997). An epidemiologic study of fatigue with relevance for the chronic fatigue syndrome. *J. Psychiatr. Res.* 31: 19.

9 Soderberg, S., Lundman, B., and Norberg, A. (2002). The meaning of fatigue and tiredness as narrated by women with fibromyalgia and healthy women. *J. Clin. Nurs.* 15: 247–255.

10 Anoymous (2021). My experience of chronic fatigue syndrome. *BMJ Paeditr. Open* 5: e001165.

11 Ho, L.Y.W., Lai, C.K.Y., and Ng, S.S.M. (2021). Contribution of sleep quality to fatigue following a stroke: a cross-sectional study. *BMC Neurol.* 21: 151.

12 Mills, R.J. and Young, C.A. (2011). The relationship between fatigue and other clinical features of multiple sclerosis. *Mult. Scler.* 17: 604.

13 Cantor, F. (2010). Central and peripheral fatigue: exemplified by multiple sclerosis and myasthenia gravis. *PM&R* 2: 399.

14 Ceban, F., Ling, S., Lui, L.M.W. et al. (2021). Fatigue and cognitive impairment in post-COVID-19 syndrome: a systematic review and met-analysis. *Brain Behav. Immun.*.

15 Kingstone, T., Taylor, A.K., O'Donnell, C.A. et al. (2020). Finding the 'right' GP: a qualitative study of the experiences of people with long-COVID. *BJGP Open* 4 (5): bjgpopen20X101143.

16 Townsend, L., Dyer, A.H., Jones, K. et al. (2020). Persistent fatigue following SARS-CoV-2 infection is common and independent of severity of initial infection. *PLoS One* 15: e0240784.

17 Van Herck, M., Goertz, Y.M., Houben-Wilke, S. et al. (2021). Severe fatigue in Long COVID: web-based quantitative follow-up study in members of online Long COVID support groups. *J. Med. Internet Res.* 23: e30274.

18 Lam, M.H., Wing, Y.K., Yu, M.W. et al. (2009). Mental morbidities and chronic fatigue in severe acute respiratory syndrome survivors: long-term follow-up. *Arch. Intern. Med.* 169: 2142–2147.

19 McIntyre, R.S., Xiao, H.X., Syeda, K. et al. (2015). The prevalence, measurement, and treatment of the cognitive dimension/domain in major depressive disorder. *CNS Drugs* 29 (7): 577–589.

20 Teodoro, T., Edwards, M.J., and Isaacs, J.D. (2018). A unifying theory for cognitive abnormalities in functional neurological disorders, fibromyalgia and chronic fatigue syndrome: systematic review. *J. Neurol. Neurosurg. Psychiatry* 89: 1308.

21 Belluck P. (2021). Cognitive rehab: One patient's painstaking path through Long Covid therapy. *The New York Times*. 3 December.

22 Groff, D., Sun, A., Ssentongo, A.E. et al. (2021). Short-term and long-term rates of postacute sequelae of SARS-CoV-2 infection. A systematic review. *JAMA Netw. Open* 4: e2128568.

23 Holson LM. (2021). My 'Long Covid' nightmare: Still sick after 6 months. *The New York Times*. 21 January.

24 Soraas, A., Bo, R., Kalleberg, K.T. et al. (2021). Self-reported memory problems 8 months after COVID-19 infection. *JAMA Netw. Open* 4 (7): e2118717.

25 Callan, C., Ladds, E., Husain, L. et al. (2022). 'I can't cope with multiple inputs': a qualitative study of the lived experience of 'brain fog' after COVID-19. *BMJ Open* 12: e056366.

26 Hampshire, A., Trender, W., Chamberlain, S.R. et al. (2021). Cognitive deficits in people who have recovered from COVID-19. *EClin. Med.* 39: 101044.

27 Becker, J.H., Lin, J.J., Doernberg, M. et al. (2021). Assessment of cognitive function in patients after COVID-19 infection. *JAMA Netw. Open* 10: E2130645.

28 Zhao, S., Shibata, K., Hellyer, P.J. et al. (2022). Rapid vigilance and episodic memory decrements in COVID-19 survivors. *Brain Comm.* https://doi.org/10.1093/braincomms/fcab295.

29 Aiello, E.N., Fiabane, E., Manera, M.R. et al. (2021). Episodic long-term memory in post-infectious SARS-CoV-2 patients. *Neurol. Sci.* 1–4.

30 Whiteside, D.M., Basso, M.R., Naini, S.M. et al. (2022). Outcomes in post-acute sequelae of COVID-19 (PASC) at 6 months post-infection: cognitive functioning. *Clin. Neuroppsychol.* 1–23.

31 Wang, S.Y., Adejumo, P., See, C. et al. (2021). Characteristics of patients referred to a cardiovascular disease clinic for post-acute sequelae of SARS-CoV-2 infection. *medRxiv* https://doi.org/10.1101/2021.12.04.21267294.

32 Rinaldo, R.F., Mondoni, M., Parazzini, E.M. et al. (2021). Deconditioning as main mechanism of impaired exercise response in COVID-19 survivors. *Eur. Respir. J.* 58 (2): 2100870.

33 Skjorten, I., Ankerstjerne, O.A.W., Trebinjac, D. et al. (2021). Cardiopulmonary exercise capacity and limitations 3 months after COVID-19 hospitalisation. *Eur. Respir. J.* 58 (2): 2100996.

34 Malik, P., Patel, K., and Pinto, C. (2021). Post-acute COVID-19 syndrome (PCS) and health-related quality of life (HRQoL)- a systematic review and meta-analysis. *J. Med. Virol.*.

35 Singh, I., Joseph, P., Heerdt, P.M. et al. (2022). Persistent exertional intolerance after COVID-19. *Chest* 161: 54.

36 Novak, P., Mukerji, S.S., Alabsi, H.S. et al. (2021). Multisystem involvement in post-acute sequelae of coronavirus disease 19. *Ann. Neurol.*.

37 Dahlhamer, J., Lucas, J., Zelaya, C. et al. (2018). Prevalence of chronic pain and high-impact chronic pain among adults–United States, 2016. *MMWR Morb. Mortal. Wkly. Rep.* 67 (36): 1001–1006.

38 Smith, B., Fors, E.A., Korwisi, B. et al. (2019). The IASP classification of chronic pain for ICD-11: applicability in primary care. *Pain* 160: 83.

39 Institute of Medicine (2011). *Relieving Pain in America: A Blueprint for Transforming Prevention, Care, Education, and Research*. National Academies Press.

40 Nicholson KM. (2021). Another fight for Covid long-haulers: having their pain acknowledged. *STAT* (2 December).

41 Taquet, M., Dercon, Q., Luciano, S. et al. (2021). Incidence, co-occurrence, and evolution of long-COVID features: a 6-month retrospective cohort study of 273, 618 survivors of COVID-19. *PLoS Med.*.

42 Ursine, F., Ciaffi, J., Mancarella, L. et al. (2021). Fibromyalgia: a new facet of the post-COVID-19 syndrome spectrum? Results from a web-based survey. *RMD Open* 7: e001735.

43 Huang, S., Zhuang, W., Wang, D. et al. (2021). Persistent somatic symptom burden and sleep disturbance in patients with COVID-19 during hospitalization and after discharge: a prospective cohort study. *Med. Sci. Monit.* 27: e930447-1–e930447-12.

44 Russell, M.B. (2005). Tension-type headaches in 40-year-olds: a Danish population-based sample of 4000. *J. Headache Pain* 5: 441.

45 Schwedt, T.J. (2014). Chronic migraine. *BMJ* 348: g1416.

46 Russo, A., Silvestro, M., Tedeschi, G., and Tessitore, A. (2017). Pathophysiology of migraine: what have we learned from functional imaging? *Curr. Neurol. Neurosci. Rep.* 17: 95–107.

47 Cho, S.-J., Sohn, J.-H., Bae, J.S., and Chu, M.K. (2018). Fibromyalgia among patients with chronic migraine and chronic tension-type headache: a multicenter prospective cross-sectional study. *Headache* 58: 311–313.

48 Amoozegar, F. (2017). Depression comorbidity in migraine. *Int. Rev. Psychiatry* 29: 504–515.

49 Beghi, E., Giussani, G., Westenberg, E. et al. (2021). Acute and post-acute neurological manifestations of COVID-19: present findings, critical appraisal, and future directions. *J. Neurol.* 1–10.

50 Al-Hashel, J.Y., Abokalawa, F., Alenzi, M. et al. (2021). Coronavirus disease-19 and headache; impact on pre-existing and characteristics of de nono: a cross-sectional study. *J. Headache Pain* 22: 97.

51 Krause, A.J., Prather, A.A., Wager, T.D. et al. (2019). The pain of sleep loss: a brain characterization in humans. *J. Neurosci.* 39: 2291.

52 Prather, A.A., Janicki-Deverts, D., Hall, M.H. et al. (2015). Behaviorally assessed sleep and susceptibility to the common cold. *Sleep* 38: 1353.

53 Liu, C., Pan, W., Li, L. et al. (2021). Prevalence of depression, anxiety, and insomnia symptoms among patients with COVID-19: a meta-analysis of quality effects model. *J. Psychosom. Res.* 147: 110516.

54 Brody, J. (2021). The devastating ways depression and anxiety impact the body. *The New York Times* (4 November).

55 Mazza, M.G., De Lorenzo, R., Conte, C. et al. (2020). Anxiety and depression in COVID-19 survivors: role of inflammatory and clinical predictors. *Brain Behav. Immun.* 89: 594.

56 Houben-Wilke, S., Goertz, Y.M.J., Delbressine, J.M. et al. (2022). The impact of Long COVID-19 on mental health: observational 6-month follow-up study. *JMIR Ment. Health* 9: e33704.

57 El Sayed, S., Shokry, D., and Mohamed, G.S. (2021). Post-COVID-19 fatigue and anhedonia: a cross-sectional study and their correlation to post-recovery period. *Neuropsychopharmacol. Rep.* 41: 50.

58 Lamontagne, S.J., Winters, M., Pizzagalli, D.A. et al. (2021). Post-acute sequelae of COVID-19: evidence of mood & cognitive impairment. *Brain Behav. Immun.* 17: 100347.

59 Vannorsdall, T.D., Brigham, E., Fawzy, A. et al. (2021). Rates of cognitive dysfunction, psychiatric distress and functional decline following COVID-19. *J. Acad. Consult. Liaison Psychiatry*; S2667–2960(21)00185–3.

60 Whiteside, D.M., Nani, S.M., Basso, M.R. et al. (2022). Outcomes in post-acute sequelae of COVID-19 (PASC) at 6 months post-infection part 2: psychological functioning. *Clin. Neuropsychol.* 32.

61 Azevedo, M.N., da Silva, R.E., Vieira Passos, E.A.F. et al. (2022). Multimorbidity associated with anxiety symptomatology in post-COVID patients. *Psychiatr. Res.* 309: 114427.

62 Silvan, M., Parkin, A., Makower, S., and Greenwood, D.C. Post-COVID syndrome symptoms, functional disability and clinical severity phenotypes in hospitalized and no, n-hospitalized individuals. *J. Med. Virol.* 20921.

63 David, A.S., Wessely, S., and Pelosi, A.J. (1988). Postviral syndrome: time for a new approach. *Br. Med. J.* 296: 696.

5

Historical Perspectives, Including Chronic Fatigue Syndrome/Myalgic Encephalomyelitis and Fibromyalgia

The illnesses that most resemble long-COVID syndrome are CFS/ME and fibromyalgia. This chapter will describe the similar symptoms of these conditions to long-COVID syndrome. We will also discuss what is termed chronic Lyme disease, or PTLDS, as an example of an analogous, unexplained syndrome following an infection. We believe that there are lessons to be learned from CFS/ME, fibromyalgia, and post-treatment Lyme disease syndrome (PTLDS) with direct applicability to better understand and treat long-COVID syndrome.

Chronic Fatigue Syndrome, Myalgic Encephalomyelitis (CFS/ME)

Dr. Anthony Fauci called attention to the similarity of long COVID with CFS/ME in July 2020, "Anecdotally, there's no question that there are a considerable number of individuals who have a post-viral syndrome that really, in many respects, can incapacitate them for weeks and weeks following so-called recovery and clearing of the virus, highly suggestive of myalgic encephalomyelitis/chronic fatigue syndrome" [1]. The next case describes a patient of ours with CFS/ME.

Anna, a 47-year-old female, was in good health until a bout of the flu in 2010. After a week of fever and flu-like symptoms, she felt better and went back to work. However, she immediately relapsed, describing, "I just couldn't get my energy back. Over the next few months, I became more and more exhausted and began experiencing body aches all over. I saw my doctor and all the tests came back normal. He thought that I had a post-viral syndrome that would get better soon. But that never happened, and I have been sick for more than one year. The exhaustion is overwhelming. I can't concentrate on anything. No one gives me any answers and I'm feeling hopeless. One doctor thought that I was depressed

Unravelling Long COVID, First Edition. Don Goldenberg and Marc Dichter.

and put me on an antidepressant. Another told me that I likely had CFS but that there was no specific treatment for it. Finally, I joined a CFS support group and found a clinic that specializes in CFS. They told me that my immune system had been damaged and began me on an elimination diet and intravenous infusions to strengthen my immune system. I felt better for a few weeks and then it all came roaring back."

CFS/ME are diagnostic labels used to describe a syndrome characterized by chronic fatigue, post-exertional malaise (PEM), muscle pain, and cognitive disturbances often attributed to a possible infection but without a defined cause or pathology. Depending on the diagnostic criteria used, the prevalence varies from as little as 0.006% to as much as 3% of the population. CFS/ME is more common in women, peaking at 25 to 50 years.

Throughout the past few centuries, there were numerous descriptions of persistent, unexplained medical symptoms attributed to outbreaks of infectious diseases. After the Russian flu pandemic of 1889–1892, persistent mental and physical exhaustion were common and the medical historian, Mark Honigsbaum, wrote, "By the middle 1890s, Russian influenza was being blamed in England for everything from the suicide rate to the general sense of malaise. . . and the image of a nation of convalescents, too debilitated to work or return to daily routines, and plagued with mysterious and erratic symptoms and chronic illnesses, had become central to the period's medical and cultural iconography" [2].

The 1918–1919 influenza pandemic, dubbed the Spanish flu, was followed by persistent, unexplained central nervous system symptoms classified as parkinsonism, catatonia, and encephalitis lethargica, or so-called sleeping sickness. Common descriptions of these symptoms included "loss of muscular energy," "apathy," and "melancholia" that often persisted for years. The poet, Robert Frost, described after recovering from a severe bout of the Spanish flu, "What bones are they that rub together so unpleasantly in the middle of you in extreme emaciation? I don't know whether or not I'm strong enough to write a letter yet," [3] and a senior Red Cross official commented, "but it also left a trail of lowered vitality. . . nervous breakdown, and other sequelae, which now threaten thousands of people. It left widows and orphans and dependent old people. It has reduced many of these families to poverty and acute distress. This havoc is widespread, reaching all parts of the United States and all classes of people" [3].

During the next 50 years, episodes of similar illnesses manifested by exhaustion and mental confusion were linked to potential infectious agents, most often in health care workers, such as the one at Los Angeles County Hospital in 1934. At the time, polio was widespread but the infection rate in the hospital nurses was eight times greater than that in the community. Many of the nurses became very ill, complaining of exhaustion, generalized pain, and cognitive disturbances but none developed the characteristic neurologic manifestations of polio.

In 1950, an outbreak of similar symptoms at the Middlesex Hospital in the UK was thought to be caused by polio. However, again the symptoms and clinical course were not consistent with polio and a new medical diagnostic term, benign myalgic encephalomyelitis, was suggested. Benign indicated no deaths, myalgic described the common occurrence of diffuse muscle pain, and encephalomyelitis suggested that brain inflammation was the root of the problem. Over the next few decades there was no good evidence for either muscle or brain pathology.

Then, what we now label CFS/ME was introduced into our medical lexicon. CFS/ME was initially thought to be triggered by an infection after a cluster of cases of unexplained fatigue and other medical symptoms were reported from Lake Tahoe, Nevada in 1985 [4]. Early research suggested that the Epstein–Barr virus might be the cause of these cases, and the condition was initially called "chronic Epstein–Barr virus infection." Exhaustive studies found no such link, and CFS/ME became the preferred diagnostic label. Subsequently, CFS/ME was described throughout the US, not usually in clusters of cases but episodically and not attributed to any specific infection. Dr. Anthony Komaroff and co-investigators suggested that "CFS was a heterogeneous illness that can be triggered by multiple different genetic and environmental factors, including stress, toxins, and exogenous infectious agents" [5].

The cardinal feature of CFS, as defined by the US Centers for Disease Control and Prevention (CDC) in 1996, is severe fatigue lasting for six or more consecutive months that was not caused by any other medical condition [5]. Other CFS diagnostic criteria include at least four or more of the following eight symptoms: post-exertion malaise (PEM) lasting more than 24 hours, unrefreshing sleep, significant impairment of short-term memory or concentration, muscle pain, pain in the joints without swelling or redness, and headaches of a new type. Dr. Simon Wessely described the cognitive disturbances of CFS/ME, "The patient needs to devote more attention and energy to motor and cognitive tasks, such as exercise or mental concentration" [6]. In 2015, the Institute of Medicine redefined CFS with a new diagnostic label, systemic exertion intolerance disease (SEID) [7]. They suggested that potential physiologic markers, specifically orthostatic intolerance, be added to the diagnostic criteria.

PEM has been considered by some to be a defining feature of CFS/ME. PEM in patients with CFS/ME was associated with mental fatigue and greater sympathetic activity [8]. PEM is a common symptom in patients with long-COVID syndrome.

Fibromyalgia Syndrome

Many patients with CFS/ME also have fibromyalgia, as did Anna, whose chronic fatigue was described above. She was sent to one of us (DG) for further evaluation. Her initial examination revealed no joint swelling or inflammation and no neurologic abnormalities. She was very tender to modest palpation in many soft-tissue locations around the elbows, knees, chest wall, neck and back. Her laboratory tests were normal. I (DG) told Anna that she had fibromyalgia, which often accompanies CFS/ME. I told her about a study I did back in 1990, when we found that most patients with fibromyalgia had symptoms consistent with CFS/ME, and many of those patients thought that their illness began after an acute viral infection [9]. Fibromyalgia, like CFS and similar syndromes such as irritable bowel syndrome, is diagnosed after excluding other medical disorders that may cause the patient's symptoms [10].

The cardinal symptom of fibromyalgia is chronic, unexplained, widespread pain, severe enough to interfere with normal life activities that lasts at least three months [11]. The pain is usually present on both sides of the body as well as above and below the waist. Initially, the pain may be localized, typically to the neck and shoulders. Patients often report numbness, tingling, burning, or creeping and crawling sensations, especially in both arms and legs, but the neurologic examination is unrevealing. Patients with fibromyalgia also uniformly suffer from fatigue and sleep disturbances, and most also report cognitive disturbances, often dubbed "fibro fog." Common descriptions of symptoms include, "It feels like I always have the flu" or "When I get out of bed, it feels like a truck ran me over." Like CFS/ME, patients look well, have normal laboratory tests, and there is no clear evidence of organ damage. More specifically, the joints and muscles demonstrate no inflammation. Depending on the diagnostic criteria used, fibromyalgia affects 3% to 5% of the general population and is more common in women, peaking at 30 to 50 years.

Fibromyalgia, like CFS/ME, has been around for centuries in various guises. Indeed, fibromyalgia and CFS/ME were historically linked, first under the label neurasthenia in the 1800s. The neurologist, Dr. George Beard, described his neurasthenia patients' chronic exhaustion, muscle and joint pain, headaches, and confusion and postulated that the central nervous system (CNS) was at fault, "My own view is that the central nervous system probably undergoes slight, undetectable, morbid changes in its chemical structure and as a consequence becomes more or less impoverished in the quantity and quality of its nervous force" [12].

The patient demographics and clinical symptoms of fibromyalgia and CFS/ME overlap dramatically. Both are more common in female individuals and peak between 30 to 50 years. In addition to the cardinal symptoms of widespread pain and exhaustion, most patients report sleep, mood, cognitive disturbances, and chronic headaches. The defining feature of CFS/ME, exhaustion aggravated by physical activity (PEM), is also universally present in fibromyalgia. In both conditions, exercise increases fatigue and pain severity, suggesting similar pathologic mechanisms [13]. Both conditions often become chronic, lasting for years, and management has been symptomatic, not curative.

Fibromyalgia often occurs in conjunction with other chronic medical conditions and is present in 20% to 30% of patients with rheumatoid arthritis, systemic lupus erythematosus (SLE), and osteoarthritis. In those conditions, there is often no correlation between fibromyalgia symptoms and activity of the concurrent medical disease.

As we have discussed, fibromyalgia and CFS/ME overlap and are co-morbid with chronic headaches, chronic gastrointestinal symptoms (irritable bowel syndrome), and chronic pelvic and bladder pain as well as chronic mood and sleep disturbances. Headaches are present in more than 50% of fibromyalgia patients, including both migraine and muscular headaches.

CFS/ME, Fibromyalgia in Long-COVID

Only a few studies have applied the specific diagnostic criteria for CFS/ME or for fibromyalgia to patients with long-COVID syndrome. One-third of patients who had persistent symptoms after SARS-CoV-2 infection met the diagnostic criteria for CFS/ME six months after the initial infection [14]. More cognitive and mood disturbances were present in the CFS/ME group compared to those who did not meet the criteria for CFS/ME. There were no differences in inflammatory or immune symptoms or test results in the subset of patients who met the criteria for CFS/ME compared to those who did not. In another report, using the Institute of Medicine's revised SEID criteria for CFS, only 13% met the SEID criteria six months after hospitalization for COVID-19 [15]. Fifty-three percent of these patients had fatigue at six months but were not classified as CFS/ME because of co-morbidities or other symptoms.

The first study to look for specific symptoms of fibromyalgia after SARS-CoV-2 infection was done in Italy, and it surveyed 616 people who had SARS-CoV-2 infection, including 10% that required hospitalization [16]. Most patients (77%) in the study were women, and the mean age was 45 years. Thirty percent of these subjects fulfilled the criteria for fibromyalgia at an average of six months after SARS-CoV-2 infection. Predictors of fibromyalgia included obesity and being male. The authors forecasted a sharp rise in post-COVID-19 chronic pain, what they called "FibroCOVID."

Nine percent of patients with new symptoms after a SARS-CoV-2 infection were considered by clinicians at the Mayo Clinic to have central sensitization, which they labeled post-COVID syndrome [17]. The symptoms included widespread pain in 90% of patients, fatigue in 80%, dyspnea in 43%, and orthostatic intolerance in 38%. The authors noted the similarities of these long-COVID symptoms to those described in other post-viral syndromes and stated that if these symptoms persist for more than six months, the patients will meet criteria for CFS or fibromyalgia.

Extrapolating from more general reports of persistent symptoms in non-hospitalized long-COVID patients, most meet the diagnostic criteria for CFS/ME and fibromyalgia. For example, in an international patient survey, fatigue was present in 98% of respondents, PEM in 89%, musculoskeletal pain in 94%, and cognitive dysfunction in 85% [18]. A similar survey from long-COVID patient support group members reported fatigue in 98%, pain in 87%, and sleep disorder in 88% of respondents [19].

It is also instructive to look at symptoms consistent with CFS/ME and fibromyalgia after other coronavirus infections. In one study of 107 patients followed carefully after severe acute respiratory syndrome (SARS), a large percentage of patients had persistent fatigue and reduced quality of life at one year [20]. A group of patients with persistent fatigue and pain at an average of 19 months after SARS-CoV-1 infection had symptoms nearly identical to patients with fibromyalgia [21]. They also had similar physiological changes in their sleep EEG, including alpha wave intrusion.

Post-treatment Lyme-disease Syndrome (PTLDS)

PTLDS has many similar features to CFS/ME and fibromyalgia but is triggered by a known microbe. Most patients with acute Lyme disease recover completely if treated soon after the infection. During the acute stage of Lyme disease, the Lyme spirochete can be identified by laboratory testing, and there is usually a bulls-eye rash at the site of the tick bite. If treatment is not prompt, the Lyme bacteria may travel to internal organs, most often joints, and this subacute Lyme disease is associated with joint swelling as well as cardiac and neurologic inflammation. Antibiotics still generally work well at this later stage to knock out the remaining inflammation. However, some patients may require immunosuppressive medications like those used to treat rheumatoid arthritis. Therefore, Lyme disease in its acute stage is similar to an initial SARS-CoV-2 infection and in its subacute form may be a model for long-COVID disease with microbial-induced immune disease.

However, a controversy has raged for more than 40 years regarding what is termed chronic Lyme disease, or what most medical experts now call PTLDS, as the following case illustrates.

Vicky was diagnosed with Lyme disease after noting a circular rash with a bulls-eye center following a walk in a wooded region in the Northeast. She had a fever and felt achy for a few days, and her primary care physician started her on antibiotics. Blood tests confirmed the diagnosis of Lyme disease. Over the next week, the rash disappeared, and she felt better. However, during the next few months the generalized body achiness and tiredness returned. She then began experiencing headaches, difficulty concentrating, and her exhaustion got much worse. Repeat laboratory tests were all unremarkable. She was referred to an infectious disease specialist who found no evidence for active Lyme disease. She then saw a physician who specialized in chronic Lyme disease. An extensive evaluation, including a spinal tap, a brain MRI, and EEG, were performed. Although the results of these tests were equivocal, the specialist said that Vicky had chronic Lyme disease and placed her on a three-month course of intravenous antibiotics. During the first month of therapy, her energy improved, but after the three-month course was finished, she felt no better. Vicky continued to feel poorly for months. A rheumatologist suggested that her symptoms were most compatible with fibromyalgia and another infectious disease specialist said that she had PTLDS.

PTLDS follows a documented case of Lyme disease that, despite appropriate antibiotic treatment, results in persistent, subjective symptoms that include fatigue, widespread musculoskeletal pain, and cognitive disturbances [22]. As in CFS/ME, fibromyalgia, and long-COVID syndrome, the physical examination, laboratory tests, and tissue examinations are unremarkable, and there is no evidence of organ damage. Even the term PTLDS demonstrates the uncertainty whether this condition is best considered a disease or syndrome. As with long-COVID syndrome, a biopsychological illness model was proposed for PTLDS [23]. Patients with PTLDS have increased sensitivity to touch, pressure, sounds and light, similar to that noted in fibromyalgia patients, and similar neuroimaging findings have been noted.

Studies of patients with PTLDS, like those in long COVID, are skewed by a lack of comparison with a controlled, uninfected population. Using a controlled study design, 1135 subjects with acute Lyme disease were compared to normal controls and controls who had a tick bite but did not get Lyme disease [24]. One year after Lyme infection, persistent symptoms, including musculoskeletal pain, fatigue, and cognitive impairment, were present in 27% of all Lyme-disease patients as well as in 21% to 23% of controls who had a tick bite but no other symptoms or were healthy (Figure 5.1).

As with long COVID, it is important to make an accurate diagnosis of the initial infection. Many people who never had Lyme disease but do have unexplained chronic symptoms are misdiagnosed with chronic Lyme disease, as noted in a recent comprehensive review in which 84% of 1261 patients referred for chronic Lyme disease never had Lyme disease [25]. The most frequent diagnoses were fibromyalgia, CFS/ME, and mood disturbances, rather than Lyme disease.

What medical experts do agree upon is, like in long-COVID syndrome, there are no viable organisms that persist and cause havoc in patients with PTLDS. Yet many patients, support groups, and some physicians believe these studies are somehow flawed. There have now been five, double-blind, placebo-controlled trials demonstrating no benefit from prolonged antibiotic therapy to treat

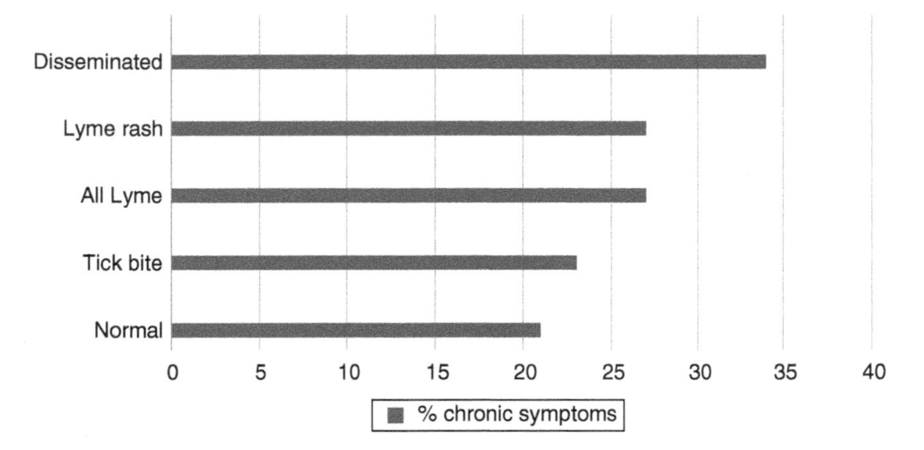

Figure 5.1 Percent of subjects with chronic symptoms at one year. *Source:* Modified from Ursinus et al. [24].

persistent chronic Lyme symptoms [26]. In one large study, 17% of patients with documented Lyme disease developed fibromyalgia, often months after they had received appropriate antibiotics [27]. The fibromyalgia symptoms did not improve with repeated courses of antibiotics. PTLDS is appropriately called a syndrome, since there is no evidence that the infection is still active, there is no well-defined organ damage, and its treatment is unclear.

Similar Patient Characteristics

CFS/ME and fibromyalgia are two to five times more common in women than in men. Inheritable and environmental factors are partially responsible for such differences. Female individuals have a heightened sensitivity to several stimuli, including pain, smell, temperature, and visual cues. These sex differences occur in humans, other mammals, birds, and fruit flies, and represent a natural, protective evolution of increased sensitivity in all species and involve hormones, such as testosterone and progesterone. Such sex hormone influences help explain why some women have attacks of migraine headaches only during their menstrual cycle and why migraine headaches often disappear with the onset of menopause.

Genetic vulnerability has been demonstrated in fibromyalgia. The first-degree relatives of patients with fibromyalgia have an eightfold greater risk of developing fibromyalgia, and family members of fibromyalgia patients have a lower pain threshold than general population controls. [28]. There is also an increased family predisposition to major depression and migraine headaches. Genetic variability interacts with sex hormone status to influence pain sensitivity. A genetic variant of an opioid receptor gene is associated with pain sensitivity in men but not in women. Neuroimaging studies have demonstrated significant gender differences in brain pain-processing regions. As discussed below, immune diseases, such as rheumatoid arthritis, SLE and multiple sclerosis (MS), are significantly more common in female individuals, and both peak between 30 to 50 years.

There are also sociocultural factors that impact gender differences in illnesses. This is apparent when one compares the gender differences in patients clinically diagnosed with fibromyalgia to individuals who meet criteria for fibromyalgia but have not been formally diagnosed. In the clinic population, there is a five to tenfold greater female prevalence of fibromyalgia, whereas in the general population that ratio is 2:1. Much of these differences are culturally driven since fibromyalgia is propagated as a female disorder, in keeping with stereotypes such as women being the weaker sex. The similar syndrome described during the 1800s, neurasthenia, was considered a result of the "nervous exhaustion" that followed the new role of women in industrialized Western society. This untoward stress caused a "depletion of the limited supply of nervous energy." Neurasthenia was considered a form of mass hysteria, a condition attributed exclusively to women. "Hysteria" comes from the Greek word for uterus.

Women also are more likely to visit a physician and more likely to report medical symptoms than men. In the clinic, fibromyalgia is considered a female-predominant illness, so physicians are biased to consider it more often in patients who are women. These cultural factors greatly increase the likelihood that a woman will be diagnosed with fibromyalgia. In the general population, the increased pain sensitivity of women may account for the twofold greater prevalence of widespread pain in women. This is an inherited trait. In the clinic, that gender difference is greatly exaggerated because of cultural factors so that fibromyalgia is five to eight times greater in female individuals. We will discuss how these same gender influences occur in long-COVID syndrome.

Shared Illness Mechanisms

Abnormalities of the central and autonomic nervous system are prominent in fibromyalgia and CFS/ME. Pre-existing sleep disturbances and a heightened stress response are strong risk factors for fibromyalgia and CFS/ME. In fibromyalgia, the most common sleep problem is a decrease of deep, nonrestorative sleep (often termed stage 4 sleep). In a prospective study of 12 000 women, self-reported sleep disturbances were the most important predictor of developing chronic, widespread pain, but other factors, such as depression, were co-contributors. The prevalence of nonrestorative sleep disturbance in the general population is approximately 11% but it is present in more than 50% of patients with fibromyalgia [11]. In a classic study from 1975, medical students slept in a laboratory while hooked up to an EEG, which monitored their brain wave activity throughout the night [29]. Whenever the EEG showed the student falling into stage 4 sleep, a buzzer sounded in the laboratory. The noise did not fully awaken the student but was sufficient to cause blips of electrical interference in the EEG, termed alpha wave intrusion into stage 4 sleep. After a few nights of this, the students complained of fatigue, generalized body pain, and were tender to palpation in specific soft-tissue spots. Sleep deprivation, as in the medical students, experimentally decreases pain thresholds to pressure, heat, and cold. Similar sleep disturbances were described in CFS/ME and in patients after SARS [30].

Alterations of the hypothalamic–pituitary–adrenal (HPA) axis and the autonomic nervous system (ANS) are prominent in both conditions. Prospective studies found that abnormal HPA axis reactivity was a strong predictor of subjects developing chronic, widespread pain. Heightened heart rate variability and other ANS reactivity were found in CFS/ME and fibromyalgia. Both disorders have been associated with POTS and small fiber neuropathy. CFS/ME patients with orthostatic intolerance during tilt table tests had a significant reduction in cerebral blood flow, which correlated with their symptoms [31]. Heart rate variability, another common ANS abnormality, is frequently present in patients with fibromyalgia.

Similar neuro-immunologic abnormalities have been reported in CFS/ME and fibromyalgia. Many of these will be discussed in Section 3, especially to compare with alterations in long COVID. Cytokines, including interleukins, interferons, and tumor necrosis factor have been elevated in patients with CFS/ME and fibromyalgia, although there is no consistent pattern. In a study of an animal model of fibromyalgia, mice were given IgG from fibromyalgia patients, which subsequently caused increased pain sensitivity in the mice and changes to pain nerve fiber density, suggesting that there is an autoimmune component to fibromyalgia [32]. Brain imaging studies in CFS/ME and fibromyalgia demonstrate an alteration in the structure and function of brain regions that are important in pain, mood, sleep, and cognition. One of the early, classic studies in fibromyalgia demonstrated that fibromyalgia patients responded to pressure with much greater pain than normal controls, and this correlated with increased activity in regions of the brain that regulate pain (Figure 5.2) [33].

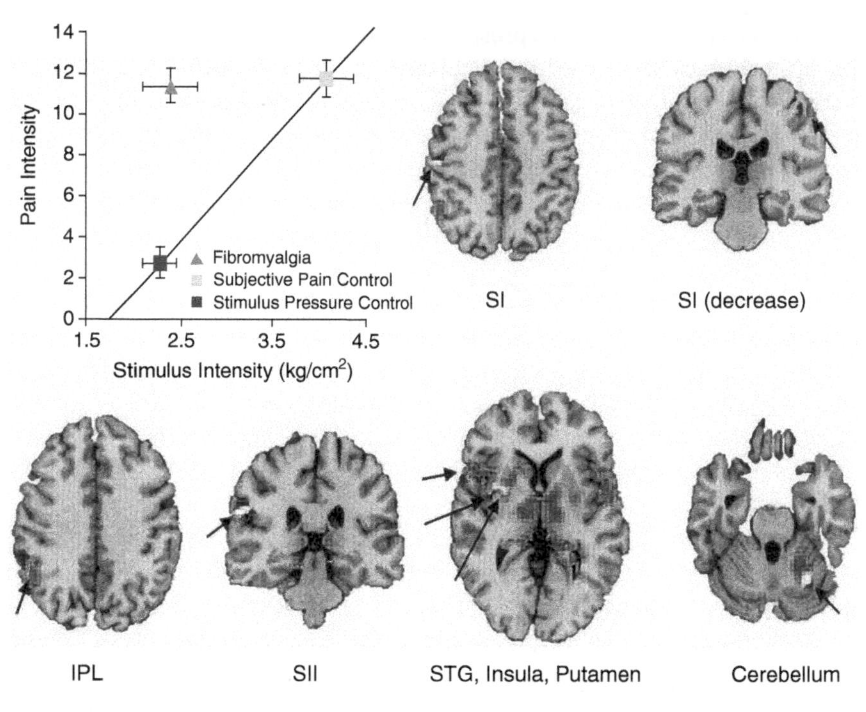

Figure 5.2 Pain intensity and brain reactivity in fibromyalgia. Top left: Pain intensity with varying pressure (stimulus intensity) in fibromyalgia patient compared to control. Imaging studies demonstrate areas of regional uptake in STG (superior temporal gyri), SI (primary somatosensory cortex), SII (secondary somatosensory cortex), IPL (inferior parietal lobe), insula, putamen, and cerebellum. *Source:* [33] Gracely et al. / reproduced with permission from John Wiley and Sons.

Lessons Learned from CFS/ME, Fibromyalgia, PTLDS

Syndrome, Disease, and Diagnostic Labels

CFS/ME and fibromyalgia are sometimes referred to as functional, somatic syndromes. There is no well-defined organ damage but there is organ dysfunction, hence the term functional. The diagnosis is based on the symptoms, rather than on physical or laboratory findings. The symptoms characteristic of CFS/ME, fibromyalgia, and long-COVID syndrome, are common in the general population. These symptoms are best envisioned on a continuum, that is, as a spectrum, based on their severity and chronicity. In contrast to a disease, there is no clear demarcation separating individuals with similar symptoms in the general population from those meeting syndromic diagnostic criteria. Many authors have commented that we all may have "fibromyalgianess" at times, just like we all get exhausted, suffer from headaches, or feel depressed at times.

When any of these symptoms become severe and persistent enough to interfere with our lives, it becomes a recognizable illness. This analogy explains why 10% to 15% of the population report chronic, widespread pain or chronic exhaustion at any point in time, but a much lower percentage (often quoted as 1% to 3%) will meet the diagnostic criteria for fibromyalgia or CFS/ME.

These illnesses are considered by many healthcare professionals to be a soft diagnosis, in contrast to objective disease diagnoses that are based on physical findings, laboratory testing, and tissue pathology. CFS/ME and fibromyalgia are diagnosed solely based on patients' subjective symptoms, "subject" to each person's individual suffering.

CFS/ME and fibromyalgia are also called invisible illnesses. Patients look well, laboratory tests and X-rays are normal, and organ damage is not present. Patients diagnosed with such syndromes may feel like second-class citizens, not worthy of the same attention as those with "real" diseases. Up until a few years ago, rheumatologists still voiced the opinion that fibromyalgia was not a "legitimate medical condition, in contrast to rheumatoid arthritis" [10]. A neurologist wrote, "Fibromyalgia is not just overdiagnosed, it downright does not exist. There is no empiricism to fibromyalgia. Just wild speculation. People with fibromyalgia do not have a specific medical disorder and they are harmed by giving them a name or treating them medically. It makes them feel crippled" [34].

Doctors find nothing objectively wrong, and patients think that their suffering is not being taken seriously. Drs. David, Wessely, and Pelosi described what often then transpired in their patients with ME/CFS, "In a desperate search for recognition, patients may resort to what one such sufferer called unacceptable patterns of behavior, which are then taken as further evidence that they are 'histrionic' or 'manipulative'. This is a manifestation of the patient's need for an acceptable diagnosis, often equalled by the pressure on the doctor to provide one" [35].

The overlapping nature of these syndromes makes physicians think they are merely "wastebasket diagnoses" or that the patient is medicalizing their symptoms, one of the "worried well." This professional skepticism and subsequent research neglect forced patients to become more vocal and involved, as we will discuss. The term long COVID was coined by patients, and described as "The first illness created through patients finding one another on Twitter" [36]. Many of the most comprehensive reports about long COVID were designed and completed by self-advocacy groups.

Patients with long-COVID syndrome are especially concerned that the comparison of their symptoms to CFS/ME trivializes their symptoms and detracts from the unique nature of long COVID. For example, one patient diagnosed with both CFS/ME and long COVID wrote, "I don't think long-haul covid should be automatically regarded as ME. Sometimes it may fit the criteria, but for the most part, I think not. All the ME and Covid people I've been in touch with says [sic] the long-haul covid feels distinctly different from the ME" [37].

CFS/ME patients and their advocates hope that the worldwide attention that long COVID has achieved will provide much-needed understanding of their own symptoms. They point out the meager prior research devoted to conditions such as CFS/ME, as noted by Terri Wilder, a patient advocate, "I've met with NIH Director Francis Collins. I've called Tony Fauci, and state senators. We still have no FDA-approved drugs, no systems of care. We only have 10–15 ME/CFS medical experts in the country. We all want our lives back, and we want this broken system fixed" [38]. Hannah Davis, another CFS/ME advocate, suggests that the limited research of conditions such as CFS/ME and fibromyalgia can still aid in understanding long COVID, "I see a lot of researchers ignoring all these fields, all the work that has already been done. They are starting from scratch. It's a huge waste of resources" [39].

Dr. Harlan Krumholz, Professor of Medicine at the Yale University School of Medicine, noted, "How many people have long-term nagging or even disabling symptoms after other viral illnesses, such as flu? We don't have a good handle on that. Maybe this is a chance for us." [39]. Dr. Walter Koroshetz, Director of the NIH National Institute of Neurological Disorders and Stroke and co-chair of the NIH's US$1.15 billion REsearching COVID to Enhance Recovery (RECOVER) research initiative to understand long COVID and similar disorders, said the pandemic is "providing scientists with a natural experiment. This is our best chance to figure out ME/CFS" [39].

Infection and Causation

Long COVID is clearly triggered by a specific virus, but it is considered a post-COVID disorder, meaning that all signs of the acute infection have abated. There are many examples of similar post-viral or other post-infectious syndromes wherein the infection triggered the syndrome but was no longer active. Infectious agents including many viruses, such as enteroviruses, human herpes virus 6, Ebola virus, West Nile virus, parvovirus, bacteria (including *Coxiella burnetii* and *Mycoplasma pneumoniae*), and parasites, such as *Giardia lamblia*, have all been associated with post-infection syndromes. One large study in Australia found that several viruses, including Ross River virus and the Epstein–Barr virus (EBV), each triggered CFS/ME in around 12% of the infected subjects [40].

In many of these post-viral syndromes, the connection to a specific infection is tenuous. When it became clear that the EBV was not the cause of CFS/ME, the diagnostic label in the US switched from chronic EBV disease to CFS. Subsequently, CFS advocates voiced concern that such a non-specific label made their condition more likely to be considered psychosomatic. Some suggested the term chronic fatigue immune dysfunction syndrome, despite the absence of any dramatic immune abnormalities. Many suggested that the term long EBV be adopted, similar to some also recommending long *Giardia*, long Ebola, or long-Lyme disease.

The diagnostic label recommended by the Institute of Medicine in 2015, SEID, was driven by patient advocacy. This was suggested "as a response to the terms CFS or BME which were unacceptable to many patients and their advocates, who reported that this term leads clinicians and others to belittle or even dismiss their disease" [7]. The new diagnostic label SEID with its updated criteria were designed to provide evidence that CFS was not a psychological illness and included the presence of physiological abnormalities, such as orthostatic intolerance.

We discussed the school of thought that chronic Lyme disease, or long-Lyme disease, is caused by persistent bacteria that can only be eradicated by intensive courses of antibiotics. This is contrary to the overwhelming evidence that the chronic pain, exhaustion, and cognitive disturbances after an appropriate short course of antibiotics are not related to ongoing Lyme infection. Most of us now use the term PTLDS, analogous to post-viral syndromes. However, the nagging concern that the Lyme infection is still active has put patients and the media at odds with healthcare providers and medical experts. Dr. Allen Steere, who discovered Lyme disease, was featured in a *New York Times* article, entitled "Stalking Doctor Steere" that described, "the growing number of patient advocacy groups and physicians who argue that chronic Lyme disease had become a full-scale epidemic, a modern-day plague crippling thousands of Americans. As the world's foremost expert on the illness, however, Steere did not believe many of them had persistent Lyme disease, but something else. . .chronic fatigue. . .or fibromyalgia and he had refused to treat them with antibiotics. . .hordes of patients had started to stalk him" [41].

Prominent voices in the media question the distinction between chronic Lyme disease and PTLDS, "There is an official consensus that regards 'post-treatment Lyme-disease syndrome' as a problem without a clear cure, and then a smaller faction of doctors who are certain that the infection itself persists and can be treated, with antibiotics and other drugs, in ways that gradually bring most patients back to health. But I want to believe that we can do better still – and that like the many people restored from Lyme with actual treatment, not just patience, there are people suffering from months and months of Covid misery who will eventually be lifted back to health" [42].

Patient-Physician Information and Misinformation

Patients with CFS/ME and fibromyalgia often express frustration with medical professionals' approach to their illness. The uncertainty regarding its root causes and the absence of organ pathology or abnormal testing are problematic for physicians schooled in classic biomedical disease models. Most discouraging, we often cannot rely on scientific studies to recommend a treatment approach.

Patients want answers and need a plan to get better. If medical professionals cannot come through, patients will seek help elsewhere. This has resulted in a more outspoken and organized patient voice in conditions such as CFS/ME and fibromyalgia. "People with fibromyalgia are either brushed off like I was or told that it is all in their head. So many of us feel betrayed by the medical profession. I was determined to find doctors who knew exactly what was wrong" [43]. When doctors cannot "find exactly what was wrong," some patients will become, "self-taught medical experts who consider themselves to be victims of a corrupt scientific establishment. . .organized its own scientific conferences, financed its own research, created its own scientific publications, and trumpeted its own medical experts" [41]. Often it is difficult for patients with CFS/ME, fibromyalgia, and PTLDS to separate fact from fiction and they may become prey to bias or downright falsehoods.

An example of the tension between CFS/ME advocates and medical professionals arose around the role of increased activity and exercise for rehabilitation. A number of scientific studies, including one called Pacing, Graded Activity, and Cognitive Behavior Therapy (PACE), published in *Lancet*, demonstrated that a carefully supervised, graded, exercise program was well-tolerated and helpful in the long-term outcome of CFS/ME patients. However, a groundswell of patient protests criticized the study and its recommendations. A CFS/ME patient wrote a widely read media piece entitled, "Bad science misled millions with chronic fatigue syndrome," which said "since patients like me were immediately skeptical, because the results contradicted the fundamental experience of our illness: The hallmark of ME/CFS is that even mild exertion can increase all the other symptoms of the disease. The researchers argued that patients like me, who felt sicker after exercise, simply hadn't built their activity up carefully enough. But I'd seen how swimming for five minutes could sometimes leave me bedbound, even if I'd swum for 10 minutes without difficulty the day before. Instead of trying to continually increase my exercise, I'd learned to focus on staying within my ever-changing limits – an approach the researchers said was all wrong" [44].

Dr. Michael Sharpe, one of the investigators in the PACE study, demonstrated that both exercise and cognitive behavioral therapy (CBT) were helpful in a significant number of patients with CFS/ME, but CFS/ME advocates strongly opposed his findings. "The Medical Research Council was being lobbied, people were trying to stop participants joining the trial – we had so much flak. One patient wrote 'As someone who is lying in bed, yet again unable to work and in agony due to long covid, I find this professor's comments so hurtful. This is a serious physical illness, not some social hysteria'" [45]. We will hear these same concerns from patients with long COVID in Section 4.

Mind or Body?

The archaic mind *or* body dualism has widened the gap between patients and providers in disorders such as CFS/ME and fibromyalgia. In the later part of the twentieth century, UK physicians concluded, "As there seems to be a total lack of objective evidence in support of the view that in cases of benign myalgic encephalomyelitis the brain and spinal cord are the site of an infective, inflammatory disease process, we would suggest that the ME name be discarded" [46].

Dr. Paul Garner, a professor of infectious disease in the UK, described his long COVID, "In mid-March, I developed covid-19. For almost seven weeks I have been through a roller coaster of ill health, extreme emotions, and utter exhaustion. Although not hospitalized, it has been frightening and long. The illness ebbs and flows, but never goes away. The symptoms changed, it was like an advent calendar, every day there was a surprise, something new. A muggy head; acutely painful calf; upset stomach; tinnitus; pins and needles; aching all over; breathlessness; dizziness; arthritis in my hands; weird sensation in the skin with synthetic materials. Gentle exercise or walking made me worse – I would feel absolutely dreadful the next day" [47].

Initially Garner rejected any analogy of long COVID to CFS/ME, stating adamantly, "The least helpful comments were from people who explained to me that I had post-viral fatigue. I knew this was wrong. I spoke to others experiencing weird symptoms, which were often discounted by those around them as anxiety, making them doubt themselves. And too, people report that their families do not believe their ever-changing symptoms, that it is psychological, it is the stress. It's deeply frustrating. A lot of people start doubting themselves, their partners wonder if there is something psychologically wrong with them. However, once I began listening to the CFS/ME community, understanding that our unconscious normal thoughts and feelings influence the symptoms we experience, provoking unconscious defense mechanisms which become established as dependent neural marks, giving false fatigue alarms. . .but could change the symptoms by retraining the bodily reactions with my conscious thoughts, feelings, and behavior" [47].

An integrated mind–body approach, what we term a biopsychological illness model, is the best way to fathom these unexplained symptoms of long-COVID syndrome. The problem with this illness model is that some physicians and many patients jump to the erroneous conclusion that we are saying "It's all in your head." Dr. Krumholz warned that any discussion that CFS/ME and long COVID are psychogenic will be seen as a rejection, "To be dismissive when the story is not yet told is shortsighted at best, cruel at worst. We are far too ignorant of the long-term effects of viruses to come to that conclusion" [39]. The CFS/ME patient/reporter who criticized the research demonstrating benefit from exercise said that such "bad science. . .has inflicted damage on millions of ME/CFS patients around the world, by promoting ineffectual and possibly harmful treatments and by feeding the idea that the illness is largely psychological" [44].

Rather, we need to discard the archaic mind/body dualism and better understand how the mind and body work in concert to orchestrate the various symptoms. Research into the cause and mechanisms of CFS/ME and fibromyalgia has been stymied by our failure to recognize the complicated brain–immune interaction. Dr. Anthony Komaroff predicted 20 years ago that, "CFS may become a paradigmatic illness that leads us away from being trapped by the rigidity of the conventional biomedical model and lead us to a fuller understanding of suffering" [48].

Summary

CFS/ME and fibromyalgia are characterized by persistent, multiple symptoms in the absence of physical or laboratory abnormalities. There is no well-defined organ damage, and they are best thought of as medical syndromes rather than diseases. They overlap dramatically and occur concurrently with other poorly understood medical conditions, such as chronic headaches, irritable bowel syndrome, and chronic bladder and pelvic pain. There are many similarities of CFS/ME and fibromyalgia to long-COVID syndrome (Table 5.1). A major concern in the near future is whether long-COVID syndrome will become a chronic illness, as CFS/ME and fibromyalgia typically do.

The primary symptoms of these syndromes, including exhaustion, widespread pain, sleep, and cognitive and mood disturbances, are the same as those of long-COVID syndrome. The mechanisms of these syndromes are unclear, but there is

Table 5.1 Similar features of CFS/ME, fibromyalgia, and long-COVID syndrome.

Considered syndromes, not diseases, with no evidence of organ damage.
Diagnosed after excluding other conditions that may cause the symptoms.
Severe enough to interfere with normal activities and last at least three to six months.
Primary symptoms include exhaustion, pain and cognitive, sleep, and mood disturbances.
More common in women, 30–50 years.
May become a chronic illness.
Infection and other stressors trigger the condition but are not active.
Pathophysiology involves brain–body interactions, CNS dysfunction, and neuroimmunity.

convincing evidence that CNS dysregulation and its interaction with the immune system is an important component. Neuroimaging studies in fibromyalgia provide evidence that, in the near future, brain signatures may provide objective diagnostic and prognostic guidelines. In Section 3, we will explore current and future research that may better elucidate these neuroimmune connections in patients with long COVID.

References

1 Rubin, R. (2020). As their numbers grow, COVID-19 "long haulers" stump experts. *JAMA* 324, https://doi.org/10.1001/jama.2020.17709: 1381–1383.
2 Honigsbaum, M. (2019). *The Pandemic Century: One Hundred Years of Panic, Hysteria, and Hubris*. London; New York: Hurst; Norton.
3 Barry, J.M. (2021). *The Great Influenza*, 392. Penguin Publishing Group. Kindle Edition.
4 Straus, S.E., Tosato, G., Armstrong, G. et al. (1985). Persisting illness and fatigue in adults with evidence of Epstein-Barr virus infection. *Ann Intern Med.* 102: 7–16.
5 Komaroff, A.L., Fagioli, L.R., Geiger, A.M. et al. (1996). An examination of the working case definition of chronic fatigue syndrome. *Am. J. Med.* 100: 56–64.
6 Wessely, S. (2001). Chronic fatigue: symptom and syndrome. *Ann. Intern. Med.* 134: 381.
7 Institute of Medicine (2015). *Beyond Myalgic Encephalomyelitis/Chronic Fatigue Syndrome: Redefining an Illness*. Washington, DC: National Academies Press.
8 Kujawski, S., Stomko, J., Hodges, L. et al. (2021). Post-exertional malaise may be related to central blood pressure, sympathetic activity and mental fatigue in chronic fatigue syndrome patients. *J. Clin. Med.* 10: 2327.
9 Buchwald, D., Goldenberg, D.L., Sullivan, J.L. et al. (1987). The "chronic, active epstein-barr virus infection" syndrome and primary fibromyalgia. *Arthritis Rheumatol.* 30: 10.
10 Goldenberg, D.L. (2019). Diagnosing fibromyalgia as a disease, an illness, a state, or a trait? *Arthritis Care Res. (Hoboken)* 71: 334–336.
11 Goldenberg, D.L. (1987). Fibromyalgia syndrome. An emerging but controversial condition. *JAMA* 257 (20): 2782–2787.
12 Beard, G. (1869). Neurasthenia or nervous exhaustion. *Boston Med. Surg. J.* 80: 217.
13 Barhorst, E.E., Boruch, A.E., Cook, D.B. et al. (2021). Pain-related post-exertional malaise in myalgic encephalomyelitis/chronic fatigue syndrome (ME/CFS) and fibromyalgia: a systematic review and three-level meta-analysis. *Pain Med.* pnab308. https://doi.org/10.1093/pm/pnab308.

14 Mantovani, E., Mariotto, S., Gabbiani, D. et al. (2021). Chronic fatigue syndrome: an emerging sequela in COVID-19 survivors? *J. Neurovirol.* 27 (4): 631–637.

15 Gonzalez-Hermosillo, J.A., Martinez-Lopez, J.P., Carrillo-Lampon, S.A. et al. (2021). Post-acute COVID-19 symptoms, a potential link with myalgic encephalomyelitis/chronic fatigue syndrome: a 6-month survey in a Mexican cohort. *Brain Sci.* 11: 760.

16 Ursini, F., Ciaffi, J., Mancarella, L. et al. (2021). Fibromyalgia: a new facet of the post-COVID-19 syndrome spectrum? Results from a web-based survey. *RMD Open* 7: e001735.

17 Bierle, D.M., Aakre, C.A., Grach, S.L. et al. (2021). Central sensitization phenotypes in post acute sequelae of SARS-CoV-2 (PASC): defining the post COVID syndrome. *J. Prim. Care Comm. Health* 12: 1–8.

18 Davis, H.E., Assaf, G.S., McCorkell, L. et al. (2021). Characterizing long COVID in an international cohort: 7 months of symptoms and their impact. *EClinMed* 38: 101019.

19 Vaes, A.W., Goertz, Y.M.J., Van Herck, M. et al. (2021). Recovery from COVID-19: a sprint or marathon? 6-month follow-up data from online long COVID-19 support group members. *ERJ Open Res.* 7 (2): 00141–02021.

20 Herridge, M.S., Cheung, A.M., Tansey, C.M. et al. (2003). One-year outcomes in survivors of the acute respiratory distress syndrome. *N. Engl. J. Med.* 348: 683–693.

21 Moldofsky, H. and Patcal, J. (2011). Chronic widespread pain, fatigue, depression and disordered sleep in chronic post-SARS syndrome; a case-controlled study. *BMC Neurol.* 37: https://doi.org/10.1186/1471-2377-11-37.

22 Rebman, A.W. and Aucott, J.N. (2020). Post-treatment lyme disease as a model for persistent symptoms in lyme disease. *Front. Med.* 7: 57.

23 Batheja, S., Nields, J.A., Landa, A. et al. (2015). Post-treatment lyme syndrome and central sensitizartion. *J. Neuropsych. Clin. Neurosci.* 25 (3): 176–186.

24 Ursinus, J., Vrijmoeth, H.D., Harms, M.G. et al. (2021). Prevalence of persistent symptoms after treatment for Lyme borreliosis: a prospective observational cohort study. *Lancet Reg. Health Eur.* 6: 100142.

25 Kobayashi, T., Higgins, Y., Melia, M. et al. (2022). Mistaken identity: many diagnoses are frequently misattributed to Lyme disease. *Am. J. Med.* 135: 503–511. https://doi.org/10.1016/j.amjmed.2021.10.040.

26 Klempner, M.S., Baker, P.J., and Shapiro, E.D. (2013). Treatment trials for post-Lyme disease symptoms revisited. *Am. J. Med.* 126: 665.

27 Goldenberg, D.L. (1993). Do infections trigger fibromyalgia? *Arthritis Rheumatol.* 36: 1489.

28 Arnold, L.M., Hudson, J.I., Hess, E.V. et al. (2004). Family study of fibromyalgia. *Arthritis Rheum.* 50: 944–952.

29 Moldofsky, H., Scarisbrick, P., England, R., and Smythe, H. (1975). Musculoskeletal symptoms and non-REM sleep disturbances in patients with "fibrositis syndrome" and healthy subjects. *Psychosom. Med.* 37: 341.

30 Moldofsky, H. and Patcai, J. (2011). Chronic widespread musculoskeletal pain, fatigue, depression and disordered sleep in chronic post-SARS syndrome: a case-controlled study. *BMC Neurol.* 11: 37.

31 van Campen, C.L.M.C., Rowe, P.C., and Visser, F.C. (2021). Cerebral blood flow remains reduced after tilt table testing in myalgic encephalomyelitis/chronic fatigue syndrome patients. *Clin. Neurophysiol. Pract.* 6: 245.

32 Goebel, A., Krock, E., Gentry, C. et al. (2021). Passive transfer of fibromyalgia symptoms from patients to mice. *J. Clin. Invest.* 131: 13.

33 Gracely, R., Petzke, F., Wolf, J.M., and Clauw, D.J. (2002). Functional magnetic resonance imaging evidence of augmented pain processing in fibromyalgia. *Arthritis Rheum.* 46: 5.

34 Bohr, T. (1996). Problems with myofascial pain syndrome and fibromyalgia syndrome. *Neurology* 46: 593.

35 David, A.S., Wessely, S., and Pelosi, A.J. (1988). Postviral fatigue syndrome: time for a new approach. *Brit. Med. J.* 296: 696.

36 Callard, F. and Perego, E. (2021). How and why patients made Long Covid. *Soc. Sci. Med.* 268: 113426.

37 Roth, P.H. and Gadebusch-Bondio, M. (2022). The contested meaning of "long COVID"–patients, doctors and the politics of subjective evidence. *Soc. Sci. Med.* 292: 114619.

38 Yong E. (2020). Long-haulers are redefining COVID-19. *The Atlantic* (13 August).

39 Sellers, F.S. (2021). Could long covid unlock clues to chronic fatigue and other poorly understood conditions? *The Washington Post* (20 November).

40 Hickie, I., Davenport, T., Wakefield, D. et al. (2006). Post-infective and chronic fatigue syndromes precipitated by viral and non-viral pathogens: prospective cohort study. *BMJ* 333: 1.

41 Grann, D. (2001). Stalking Doctor Steere. *The New York Times* (8 July).

42 Douthat, R. (2001). Long-haul Covid and the chronic illness debate. *The New York Times* (2 February).

43 Goldenberg, D.L. and Miller, A.M. (1999). Fibromyalgia on the internet: a misinformation superhighway. *Arthritis Rheumatol.* 42: S151.

44 Rehmeyer J. (2016). Bad science misled millions with chronic fatigue syndrome. Here's how we fought back. *STAT*. (21 Aug).

45 Newman, M. (2021). Chronic fatigue syndrome and long covid: moving beyond the controversy. *BMJ* 373: n1559.

46 McEvedy, C.P. and Beard, A.W. (1970). Concept of benign myalgic encephalomyelitis. *Br. Med. J.* 1: 11–15.

47 Garner, P. (2020). For 7 weeks I have been through a roller coaster of ill health, extreme emotions, and utter exhaustion. *BMJ*. May 5, 2020.

48 Komaroff, A.L. (2000). The biology of chronic fatigue syndrome. *Am. J. Med.* 108: 169.

Section 3

Mechanisms and Pathways

6

Brain Homeostasis Run Amok

This chapter will elucidate how normal homeostatic brain mechanisms may run amok (go awry) and how this process explains the presently unexplained symptoms of long-COVID syndrome. We will describe how reorganizations in brain connectivity, referred to as neuroplasticity, provide us with a framework for the brain independently perpetuating symptoms such as fatigue, pain, cognitive problems, sleep disturbances, and headaches in the absence of peripheral organ damage, and review supporting studies done in CFS/ME and fibromyalgia. We will also highlight evidence from recent reports that such neuroplasticity is associated with the persistent symptoms of long-COVID syndrome.

Brain Homeostasis in Health and Disease

To understand how long-COVID syndrome symptoms occur in the absence of organ damage, it is necessary to review a few fundamental principles about how our brains work and generate our thoughts and feelings as they keep the body's functions working properly–the state called homeostasis.

The human nervous system is a complex system of elements that control our bodily functions, including those we think about and those that occur automatically, without our awareness. We are conscious of our brain at work as we sense the external world via touch, sight, hearing, smell, and taste. We are also aware of our brain when it controls our actions and thoughts. Behind these sensory, motor, and cognitive functions, the brain and the rest of our nervous system perform a large series of complex computations that allow the smooth function and coordination of these functions. The brain has a difficult task managing all that it must do to keep us functioning properly. Attesting to all the work the brain must do to maintain body homeostasis, the brain, which in an adult is about 3% to 4% of the

Unravelling Long COVID, First Edition. Don Goldenberg and Marc Dichter.

body's mass, is responsible for about 20% of the body's oxygen consumption and the basal metabolic rate (BMR). By contrast, the heart of an average person may beat more than 2.5 billion times during their lifetime without any rest but uses only 1% of the body's oxygen and 7% of the body's BMR.

The brain's automatic responses, those not engaged in consciousness, work via the autonomic nervous system (ANS), which controls breathing, heart rate, blood pressure, hormone secretions, and bowel functions, among other things. The ANS part of the brain is a key homeostasis support system that governs the biological rhythms in our lives, such as sleep–wake cycles, circadian rhythms, hormonal rhythms, and so on. The ANS and many of our hormonal systems constantly sense our internal functions and adjust them to maintain the body's homeostasis.

Our conscious brains control how we perceive the world, mostly accurately, but sometimes less so, depending on the state of our inner self (for example, our emotional state, general health, and the proper functioning of our internal sensors). Normally, we remember things that we perceive and react rationally or emotionally to our environment. However, at times our internal emotional state can affect our reaction(s) to external factors. For example, if we are depressed, we do not feel the pleasures of experiences that would be very satisfying under more normal circumstances. If we are very distracted, say in the midst of a chaotic and dangerous situation, we may ignore a sudden, painful wound. How we react to stress and pain are particularly related to our baseline emotional state when we experience them.

Our Physical Brain Regulates Our Thoughts

We go through life constantly thinking. We may react to an external stimulus in a simple way, such as when we prick our finger on a rose thorn, feel momentary pain, and say "ouch." We also react to a more sophisticated sensory input, such as when someone seems to be laughing at us and begin to wonder what we might be doing, or look like, to have provoked that laughter. In the latter case, our thinking could be far away from the situational reality. Perhaps the person was laughing at a joke told by their companion. We also daydream about things that happened in the past or that we would like to happen in the future. In these cases, our thoughts are formed by the physical, approximately 3-pound, mass of protoplasm in our heads that constructs these thoughts in ways that scientists do not clearly understand.

Careful scientific studies revealed that the electrical activity of brain circuits involved in many conscious processes, including sensory perceptions, occurs before we are aware of the stimulus or actively decide to perform responsive tasks. This begs the question of whether the famous seventeenth century philosopher and mathematician, Rene Descartes's, declaration "I think, therefore, I am" should be reworded to "I am, therefore, I think"! That is, the physical, biological process that sets much of our behavior in motion is generated in the brain before we consciously experience a stimulus; before we decide to act. The time difference between the brain signals and the conscious perception of an event, or our decision to act is very small, and we do not consciously experience the delay or recognize the order of the process.

This is demonstrated in a psychological experiment where people are exposed to a very frightening scene. Monitoring instruments show that pupils dilate, heart beats accelerate, breathing increases, palms sweat, and muscles tense, all occurring just before the person indicates feeling frightened. The rapid reaction of our ANS acts quickly to a threatening stimulus before we consciously perceive the threat.

The way we think about our world and the state of our bodies are generated by our brain, and these thoughts may not be accurate if the networks that generate our feelings are not working normally. Yet, we may act on those thoughts regardless and, at least temporarily, believe they are accurate. It is possible for us to feel things that are not there and act inappropriately on that false information. These feelings may be concrete feelings like pain, be emotions like fear or depression, or be threats that are not real, and so on. Disruptions in the construction of our thoughts can also cause confusion and the inability to mentally function as normal.

Brain Circuits and Networks

The human nervous system consists of many, highly interconnected circuits or networks that control everything the body does, from simple reflexes to higher-order thinking, and everything in between. These circuits control the things we experience by way of our sense organs (eyes, ears, skin, nose, tongue, etc). The networks also control our physical activities, such as walking, playing sports, exercising, and speaking. These circuits, in an even more complex way, control how we perceive the world, how we react emotionally, how we think, and how we feel.

A simple metaphor that compares one aspect of the brain to what we know about computers may illustrate an important point. If we consider the simple interaction of two items – a silicon computer chip with transistor circuits and a program (or thought) developed by a human. Neither of these accomplish anything alone, but together they can perform several functions and be very powerful! Now, what happens if there is a typo in one line of the program code, or a transistor on the chip becomes unstable? If the change is subtle, the system fails to produce the correct output even if it seems to be working it is producing incorrect information! If the computer worked flawlessly for months before the error occurred, you may have so much faith in the output that you will believe it even when small error makes the computer's hidden calculations incorrect. You may even act on the assumption of the computer's accuracy. Thus, you might develop a false sense of reality because the tool you are confidently using now provides wrong information. Fortunately, in the case of the computer, once the output error is noted, the chip or the program can be fixed so that the two pieces again work harmoniously and produce accurate results. This is not so easily done when part of the network in the brain is disrupted.

Neuroscientists, who study the organization of the human brain, discovered long ago that different parts of the brain each receive and process different sensory information. They have traced the pathways from our peripheral sense organs through the spinal cord and up through brain circuits to the parts of the brain that feel sensation. They also map the circuits that allow us to move our bodies in a carefully regulated manner. These pathways are not straight lines but contain branch points to distribute the different parts of our experiences to different areas of the brain. Once the memories are stored, there must be a way to retrieve these parts and bind the various components into a coherent memory. Even simple voluntary movements *require* signals to be distributed to multiple brain areas.

There are multiple ways to analyze these networks using current brain imaging and electrophysiological techniques. A routine MRI can detect acute or chronic loss of brain tissue, such as after a stroke or in Alzheimer's disease. With advanced brain imaging, we can now look at brain structure in fine detail with quantitative analyses and examine aspects of brain function.

This is a particularly useful observation in the rare cases where individuals have several MRIs while healthy (such as in the UK Biobank study) that can then be directly compared to an MRI taken during an illness. In most cases, however, there is no prior MRI and the MRIs taken during an illness are compared to age-matched images.

Scientists can now monitor the intactness and function of brain networks during tasks using several different but complementary techniques. A functional MRI (fMRI) gauges a physical change in the brain's structure as well as its connectivity because when an area of the brain is active, it receives a slightly higher amount of oxygenated blood flow. This becomes important for functional diseases, including CFS/ME, as we will review. These are conditions manifested by abnormal brain function, not necessarily accompanied by structural damage.

We can then trace, for example, the network involved in pain perception, and determine if patients with chronic pain and no determined source use the same neural network as a group of individuals experiencing only a brief painful stimulus. It is possible to compare pain networks when they were functioning normally, such as in response to an acute painful stimulus and later, when they developed a chronic pain syndrome. A similar MRI technique can also show the neuronal fiber tracts that spread out in the brain and indicate if one, or more, network re-routing or disruption might explain their illness.

Another form of imaging called positron emission tomography (PET scan) uses a very small amount of radioactively labeled molecules to analyze other brain functions. One type of PET scan measures the metabolic energy that different parts of the brain use at any specific time. This can tell us if some areas are not using the normal amount of energy while others use more than normal. Identifying these regions may allow them to be correlated with the patient's symptoms and help with a diagnosis.

PET scans examine the physiological and biochemical components of specific brain networks by tracking radioactively labeled molecules as they are taken up by nerve cells or bound to cellular receptors on specific brain cells. For example, PET scans measure the amount of an important neurotransmitter, dopamine, in areas of the brain that control coordinated motor movement in patients with Parkinson's disease. PET scans and magnetic resonance spectroscopy can be combined with MRI to determine the levels of several neurotransmitters important in pain, mood, and energy such as glutamine, serotonin, and gamma amino butyric acid (GABA).

Finally, we can measure the electrical activity from the surface of the brain with electroencephalograms (EEGs), which are especially useful in elucidating brain activity during sleep and sleep abnormalities. Tracing pathway circuits with this technique is less developed than with some imaging techniques, but both EEGs and fMRIs can demonstrate default mode states where brain regions that are activated together appear in the brain when we are trying to rest peacefully. Analyzing the intactness of the default modes, or their disorganization, can provide clues as to the causes of several otherwise unexplained brain disorders.

Neuroplasticity

Our brains are highly changeable organs. In fact, they are changing as you read this chapter. Something happens to the physical structure of the brain that allows you to remember what you are reading and to integrate that new knowledge into a lot of prior experiences that were also preserved by earlier brain changes. This is not a simple issue. Each memory, for example, has multiple components. And when the retrieval of that memory occurs all those components need to be bound together. For example, if you meet someone at a party, you may remember their name (symbol), face (visual image), the sound of their voice (auditory input), whether they wore a strong perfume or cologne (olfactory sensation), whether you had a sensation of attraction, or possibly revulsion (emotional response), and even whether they were important for your career or a famous public figure (salience, or of personal importance). Each of these parts of the memory developed in different brain networks and may be stored in different brain regions. But all must be recalled and knit together when remembering them.

The technical word for this ability of the brain to conduct these steps for each experience we have is neuroplasticity. The changes in brain structure due to neuroplasticity can occur in multiple ways and for multiple causes. Some changes are clearly beneficial, such as remembering what you are reading now, a conversation you had yesterday, or a major social event 10 years ago. Other components of neuroplasticity may have a negative effect on brain function, such as remembering a stressful or traumatic event. Changes in either direction can be very small or, in some cases, large, like after a brain injury. Every experience immediately, physically alters something in our brain. Forming memories is daily neuroplasticity. Forgetting some of these memories is equally a form of neuroplasticity. Many neuroscientists throughout the world study memory formation and erasure in multiple animal species, from primitive worms to flies, mice, and rats.

Many elements of the human brain circuit formation are present at birth, whereas others appear during the early years of human development after birth. For example, simple sensory motor reflexes are hard wired into the early brain and other parts of the nervous system. Very young infants withdraw a limb from a painful stimulus just as adults do.

A pin prick, for example, activates a pain receptor in the skin that stimulates nearby sensory nerve fibers to transmit that information up the line. First, the sensory nerve cell fibers (axons) transmit the signal to the spinal cord. These axons then stimulate two kinds of next level neurons. One such group of cells, motor neurons, send a signal to the relevant limb muscles to direct a rapid, coordinated contraction and withdrawal of the limb from the stimulus. The second connection goes to other spinal cord neurons that are inhibitory–that is their signals stop neurons from firing. These inhibitory neurons communicate with the motor neurons of antagonizing limb muscles that might have blocked the withdrawal and tells them to stop. Thus, the limb produces a very quick and safe rapid movement away from the stimulus, automatically.

But that is not the whole story. While all this occurs in a fraction of a second, the infant has not yet felt pain. Pain is felt when the pain signal travels up the spinal cord into the part of the brain that consciously senses the pain. Thus, it takes a slightly longer fraction of a second for the baby to feel the pain and cry out. Furthermore, if the baby was already wet and in need of a change, or hungry, that response may be amplified. If the baby had just been fed and was very calm and peaceful, the cry might be very brief if the stimulus was quickly removed.

The point is that some complex, multi-nodal connections are built into our brains and peripheral neuronal circuits. These are present at birth and likely active immediately. Another point is that this response could be disrupted by any small problem in each of the parts.

In addition to simple reflex reactions, very young babies have built in circuits for more complex responses. These include the ability to mimic facial changes, like sticking out their tongue in response to others, and even, at a very early age, recognize and imitate behaviors like helping others that help them.

This is referred to as developmental neuroplasticity and relates to the new things we acquire in the early years of life. These can be skills like learning to walk, feed oneself, speak one or more languages, and so on. We have the hardware to learn these skills, but the brain must acquire new information to use the hardware properly as well as how to use trial and error to correct things that do not work out well. Thus, some babies learn to speak English while others may learn to speak Chinese, French, or any other language to which they are exposed, or even multiple languages at the same time.

Brain Dysfunction

What happens when our brains respond to injury, infection, or stress? This is reparative neuroplasticity and relates to the brain's reaction to network disruptions or injuries. These repair mechanisms are not fool proof, but often restore relatively normal function. At other times, however, the repairs, that is, their reorganization, go awry. When brain cells are damaged, a set of new brain circuits are made, presumably to restore normal function but such new circuitry may make the situation worse.

A good example of this might occur when someone with a brain injury develops epilepsy. Researchers hypothesize that the brain attempts to recover from the injury with new connections to compensate for the injured or lost nerve cells but instead produces a set of brain circuits that are too excitable (Figure 6.1). The reorganization of the brain network, with new synapses and circuits, causes periodic, unpredictable, organized electrical discharges called seizures (Figure 6.1).

This was demonstrated in animal models by tracing the changes in cellular anatomy and connectivity in brain areas that produce the seizures, not only in the directly injured regions but in others as well. The seizures may be relatively small and only involve small areas of brain or may progress to convulsive events if they spread to the entire brain. In many situations, the seizures themselves may produce more problems than the original brain injury. This is clearly a response to detectable physical damage of the brain. An analogous, less easily detected, change in network properties can also occur after more subtle brain disturbances, as we see in both long-COVID syndrome and in patients with other, non-COVID-related disorders, such as fibromyalgia and CFS. In these individuals, the initial precipitating event does not appear on routine, structural CT or MRI scans, or even pathological examinations of the brain.

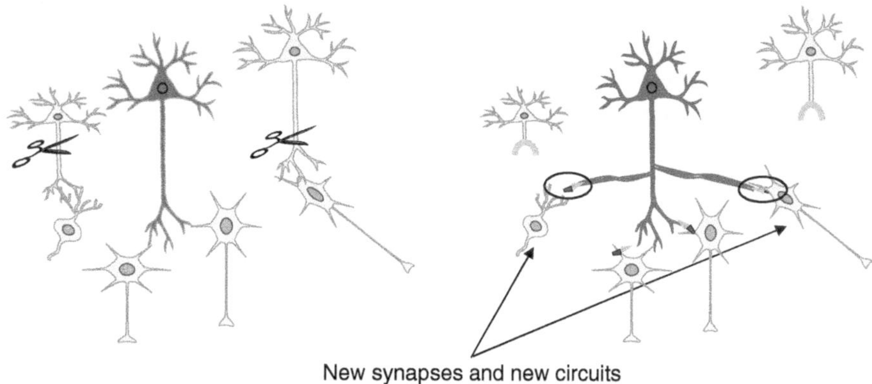

Figure 6.1 Schematic diagram of one response to brain injury that involves reorganized synaptic connections in a brain network. If axons are cut (left) they sprout and innervate nearby neurons that have lost their connections (right). Many other alterations in brain networks also occur as the brain tries to repair itself, as postulated to occur in long-COVID syndrome.

Aberrant brain network reorganization and neuroplasticity at a different level can explain why so many individuals with long COVID, fibromyalgia, or CFS may feel ill despite extensive testing of heart, lungs, and muscle strength that demonstrate normal function. One may often experience chronic pain without an external cause of the pain, shortness of breath with normal lung function, or excessive fatigue with normal muscle function. When such brain network reorganization becomes persistent, it is difficult to comprehend for both patients and physicians.

Furthermore, when one such network is disturbed, the behavioral effects spread, and other brain networks may become affected. As discussed, individuals experiencing chronic pain frequently become depressed, develop sleep disturbances, fatigue, and experience memory and cognitive problems. These symptoms occur together in the absence of external factors, following the same network disruptions. The co-occurrence of these multiple symptoms is a characteristic feature of long-COVID syndrome. Typically, the symptoms increase and decrease in tandem, so if the fatigue increases, so do the pain, headaches, sleep disturbance, depression, anxiety, and cognitive dysfunction. The same patterns seen in long COVID are found in those who suffer from chronic fatigue and chronic pain. The brain has independently created a complicated pattern of disruptive symptoms [1].

These slight detours in our brain circuits may make us think that we have a problem in our body when none exists. One of the most striking examples is phantom limb syndrome, originally described by Ambroise Pare in the sixteenth century. Between 50% and 80% of amputees suffer from severe pain at the site of an amputated limb. Phantom limb syndrome illustrates the interplay between peripheral pain, damaged nerve fibers at the amputated stump, and central pain, meaning reorganization in the pain pathways of the brain. This central pain reaction is visible on brain imaging, which demonstrates alterations in networks of the primary somatosensory cortex [2]. Individuals with phantom limb syndrome and other poorly understood pain disorders also experience an expanded field of pain as well as a generalized heightened sensitivity to light touch, noises, smells, and other non-painful stimuli.

The physical changes in the brain produced by this network reorganization syndrome are visible when areas of the parietal cortex (where the sensation of pain originates) that were originally mapped to the now-missing hand (for example), become occupied by sensory signals of the face, as the adjacent facial area spreads to replace the former hand area. When this happens, a gentle touch on the face may produce a sensation of pain in the missing hand. This all relates to changes in the brain's pain network. Eventually, the persistent pain becomes hard wired, into the remodeled brain and spinal cord.

Detecting Structural and Functional Brain Changes

In Chapter 3, we reviewed long-COVID disease, which occurs in patients with clear brain damage acquired during the initial COVID-19 illness. In this chapter, we discuss patients who do not have obvious brain damage but clear physiological and anatomical disturbances in the CNS and ANS that may cause their symptoms. In that regard, it is important to distinguish readily detected brain damage from brain dysfunction.

Brain damage, as may occur from strokes, hypoxia, or traumatic brain injuries, are detected by brain imaging or more sophisticated neurologic tests. However, more subtle forms of autoimmune neurological/psychiatric disorders cause brain dysfunction and network reorganization without the gross pathology commonly seen in brain damage. As discussed in Chapter 7, the differences between brain damage and brain dysfunction become blurred.

Brain imaging tools such as MRI, as discussed in phantom limb studies, provide a window through which to document the structural and functional brain reorganization that we believe explains many symptoms of long COVID. They have proven especially illuminating in our understanding of chronic pain conditions, such as fibromyalgia and low back pain. Fibromyalgia is a prototype for central sensitization, demonstrated by its heightened generalized sensitivity to pressure pain as well as sounds and smells [3]. Brain imaging studies completed 20 years ago showed an increased neuronal activation in the pain-processing brain areas of fibromyalgia patients compared to controls when given the same painful stimulus (Figure 5.2) [4].

In fibromyalgia and chronic low back pain, there is a loss of brain gray matter, and such structural changes are proportional to the degree and duration of pain. These structural changes are most prominent in brain regions that transmit pain, such as the insula, prefrontal cortex, thalamus, and amygdala, the pain matrix. If the chronic pain is alleviated, such structural changes often normalize.

Functional brain studies (fMRI) have documented significant differences in the connectivity of brain regions involved in pain, like the insula, and neurochemicals that promote pain transmission, including glutamate [3, 4]. A number of structural and functional abnormalities have been described in CFS/ME, some of which correlate with the extent of fatigue or sleep disturbances [5]. Changes in the hippocampus and cingulate cortex are notable. By combining such brain imaging, researchers can draw a map of chronic pain rewiring called a neural signature that may be used to explore risk factors and treatment response in the future [6].

The connectivity of various brain regions is an interest in chronic fatigue and depression research. For example, brain imaging shows that people with severe depression have a thinner hippocampus, the part of the brain that deals with learning, memory, emotions, hormonal control, and the ability to navigate the environment, among other things. When the depression subsides, some of this lost brain mass is restored by poorly understood processes. Functional imaging studies in major depression identified reduced activity in the dorsal anterior cingulate cortex that correlated with cognitive disturbances [7].

At this early stage, there is some preliminary evidence that brain reorganization may be responsible for many of symptoms of long-COVID syndrome. As discussed in Chapter 3, the loss of smell and taste, so characteristic of long COVID, is an example of organ damage. However, the initial damage to the odor-bearing receptors that stimulate olfactory nerve terminals may have triggered independent brain dysfunction. In that regard, the UK Biobank study is intriguing. All the UK Biobank participants were scanned one or more times over several years before the COVID-19 pandemic, so their "normal" or pre-infection scans were available for comparison [8]. These investigators compared MRI scans from 395 confirmed COVID-19 cases (only 15 patients had been hospitalized) to scans from 388 controls, matched for age, sex, and ethnicity. They found a loss of brain tissue in the known olfactory pathway of COVID-19 patients. This pathway includes brain regions that are functionally connected including "the piriform cortex, parahippocampal gyrus/perirhinal cortex, entorhinal cortex, amygdala, insula, frontal/parietal operculum, medial and lateral orbitofrontal cortex, hippocampus, and basal ganglia" [8]. Interestingly, the differences were predominantly confined to the left hemisphere.

There were no clear differences between the 15 hospitalized patients and those who were not hospitalized, but the numbers were too small for significant findings. The group's focus was initially on the olfactory pathways. However, their findings of reduced brain gray matter (the part of the cortex that houses most neurons) illustrated that critical brain areas for memory, decision making, language, emotional behavior, and depression were all involved. Were it not for the pre-COVID MRIs accumulated by the UK Biobank from those who later experienced the loss of smell, these individuals would likely have been deemed "without organ damage in the brain" from their COVID-19 illness. This study demonstrates that the brain dysfunction after SARS-CoV-2 infection was not related to a preexisting brain vulnerability.

One year after having COVID-19, patients had significant differences in the resting state functional connectivity of brain regions that regulate mood and sleep compared to controls [9]. COVID-19 patients without initial brain lesions on conventional MRI, had decreased cortical blood flow and abnormal cortical thickness as well as changes in white matter structure compared to healthy controls at three months [10]. An uncontrolled report of subjects who recovered from COVID-19 without obvious brain disease found both structural changes and decreased network connectivity three months after infection, and these changes correlated with mood disturbances [11]. SARS-CoV-2 infection left molecular signatures in the brain of recovered patients similar to that of normal aging [12].

Imaging studies using [18F] FDG PET/CT uptake found alterations in brain metabolism three to four months after COVID-19, compared to controls [13, 14]. Metabolism was decreased in areas of the brain involved in pain, smell, cognition, and headaches. Hypometabolism in the frontal lobe correlated with cognitive deficits in long COVID [15]. Cognitive impairment also correlated with altered brain glucose metabolism in subjects with long COVID [16]. These studies did not find increased brain metabolism that would be expected in areas of brain inflammation. A study looking at neuropsychological and neurophysiological parameters found evidence for central neuromotor and cognitive fatigue, apathy, and executive dysfunction; the authors suggested that a cortical impairment of inhibitory GABAergic neurotransmission could underlie these findings [17].

A group of COVID-19 patients, age 50 to 70 years, underwent several brain imaging techniques that examined gray matter morphology, cerebral blood flow (CBF), and the volume of white matter tracts at 3 months and 10 months after hospital discharge [18]. None of the patients had neurological symptoms initially or at three months, and this was the first study to explore dynamic brain changes after COVID-19 in patients without clear neurologic symptoms. The patients were grouped into mild or severe disease and the imaging studies compared to normal controls. The authors noted several structural changes, including multiple areas of reduced gray matter volume, reduced blood flow, and reduced white matter volume in both mildly and severely affected patients, compared to age-matched control patients. Many of the changes were in critical brain regions. The authors did not indicate whether there were cognitive or behavioral accompaniments to these altered structures, but it would be surprising if there were none.

The predominance of older male patients in this study is not representative of long-COVID syndrome. Nevertheless, these data provide objective neuroimaging evidence for extensive brain abnormalities in hospitalized patients, even in those without obvious neurologic symptoms. These included shrinking brain areas, white matter shrinkage, and low CBF in multiple areas. Perhaps, more importantly, the data also demonstrate that many of these conditions recovered by 10 months post-discharge, although some remained problematic in the more severely affected patients. This suggests that while potential network disruptions and the resultant shrinkage of brain tissue occurs in long COVID, it may be reversible (Figure 6.2).

The UK Biobank investigators went on to determine if their data could identify brain functional connectivity networks in COVID-19 patients that might predict a higher risk-taking behavioral signal and whether that might correlate with individuals who later developed COVID-19 (before the very transmissible omicron variant). An average of three years before COVID-19 infection, they examined MRI connectivity patterns that were more common in COVID-19 patients to predict risk factors for infection and persistent symptoms. "The predictive models successfully identified six fingerprints that were associated with COVID-19 positive, compared to negative status (all p values < 0.005). Overall, lower integration across the brain modules and increased segregation, as reflected by internal,

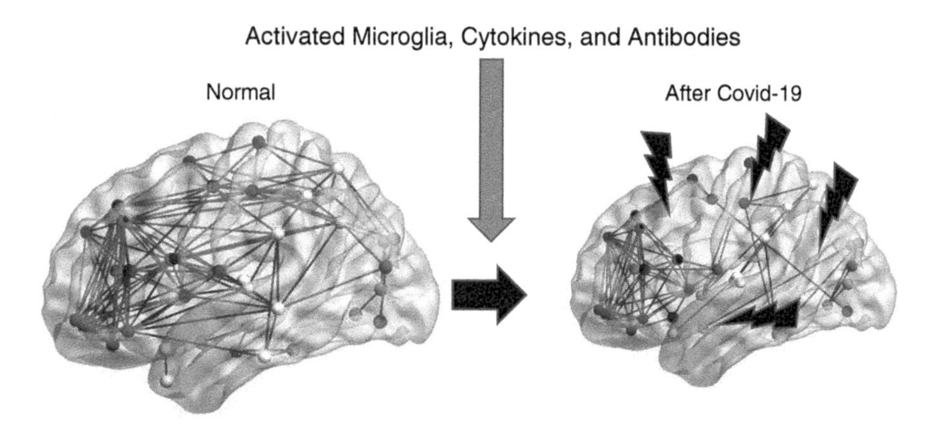

Figure 6.2 Schematic illustration of the reduced and reordered brain networks along with diminished gray matter seen with fMRI imaging in the brains of patients with long-COVID syndrome. Lightning bolt symbols indicate areas of potential gray matter loss.

within module, connectivity, were associated with higher infection rates." They concluded, "individuals are at increased risk of COVID-19 infections if their brain connectome is consistent with reduced connectivity in the top-down attention and executive networks, along with increased internal connectivity in the introspective and instinctive networks" [19].

It is possible that such brain signatures identify individuals more likely to suffer from long COVID, similar to what has been previously reported for subjects with fibromyalgia. Using fMRI to provide reliable information about behavioral characteristics is a very new and, perhaps, still controversial field; these data will need additional independent confirmations.

Changes in the Autonomic Nervous System, Neurohormones, and the Stress Response

The ANS is a major component of the overall nervous system that is partly embedded in the brain itself, but it is also comprised of multiple components outside our brains and throughout our bodies, in collections of neurons called sympathetic and parasympathetic ganglia. Within the brain, the ANS controls breathing, alertness, the sleep–wake cycle, cardiac function, gastrointestinal function, and other body functions. The peripheral ganglia have nerves that connect to multiple organs such as the heart, blood vessels, lungs, gut, kidneys, and so on (Figure 6.3). The ANS has its own set of sensors, such as those that monitor blood pressure, fluid volume, breathing, gut motility, and cardiac functions. The ANS is also comprised of endocrine organs, such as the adrenal gland, pituitary gland, and pancreas, which respond to stimuli by secreting hormones that regulate many body functions, often without our conscious awareness of their efforts.

The ANS is a critical part of the brain that responds to injury, infection, or stress. Like other components of the nervous system, it can go off kilter in response to excessive stress. The amygdala, our brains' fear center, activates the ANS and initiates the fight-or-flight response. The amygdala directly connects to autonomic centers in the hypothalamus and brainstem, demonstrating the bidirectional dysfunction that may arise in the CNS and ANS. The ANS works together with our neuroendocrine system to trigger our response to stress. The hypothalamus relays signals to the pituitary gland and adrenal glands, which release neurohormones like epinephrine. Almost immediately, our pupils dilate, hairs on the skin rise, palms become sweaty, blood pressure and heart rate increase, and our attention becomes focused, all part of a normal response to acute stress.

Prolonged stress and persistent ANS hyperactivity, however, can cause brain dysfunction due to anatomical changes in the brain. These issues relate to differentiating brain dysfunction from anatomical structural damage. This is

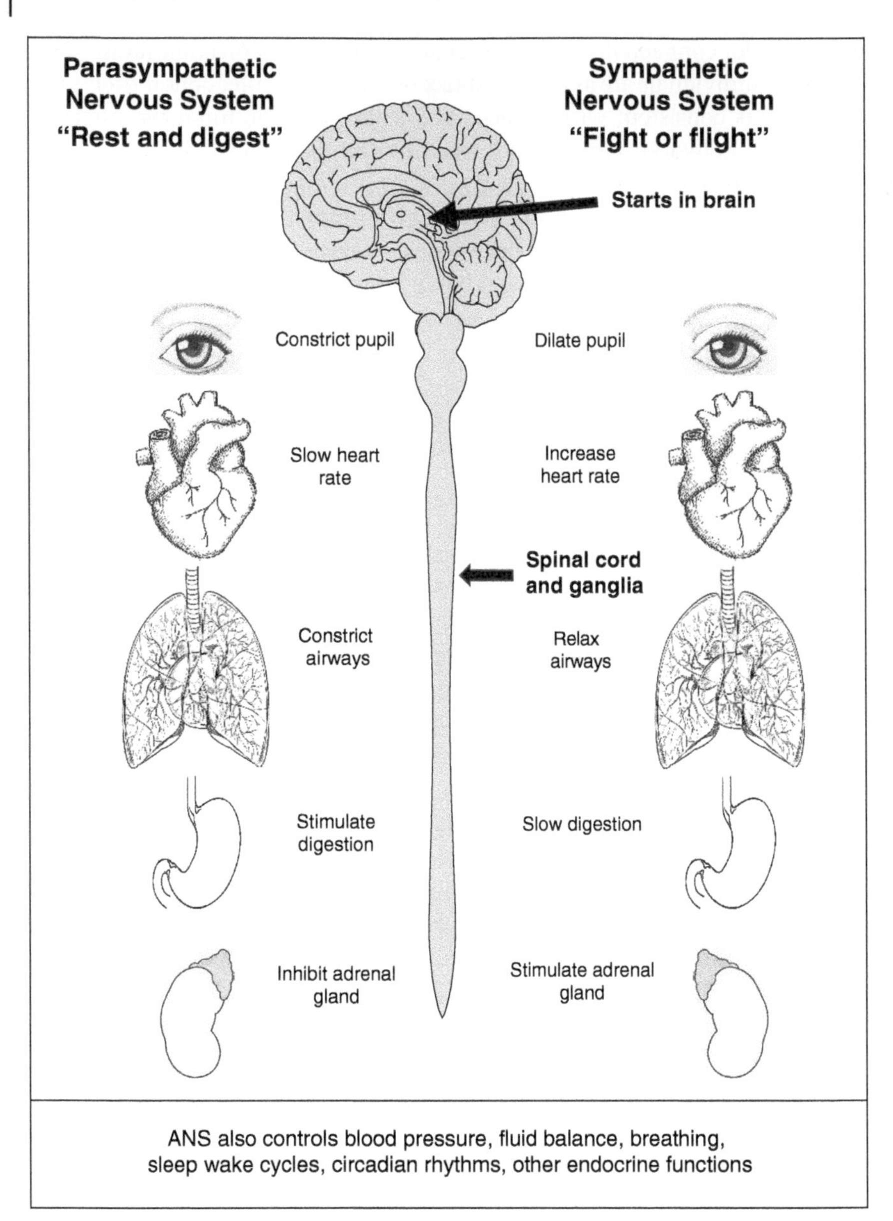

Figure 6.3 Autonomic nervous system cartoon (partial). ANS also controls blood pressure, fluid balance, breathing, sleep–wake cycles, circadian rhythms, and other endocrine functions.

particularly difficult when some damage or dysfunction can be reversible. For example, structural changes can be reversed in some cases, such as when the loss of volume in the hippocampus in severely depressed individuals is reduced if the depression is effectively treated. These issues will be discussed in more detail in Chapter 7 on neuroimmunological mechanisms relating to the long-COVID syndrome.

Hormones called glucocorticoids, secreted from the adrenal gland, can damage the hippocampus and cause it to shrink over time. Even relatively low levels of stress, especially if prolonged, can produce harmful effects. The physiological effects of this stress can be measured by an analysis of the brain networks involved as well as the hormonal changes, if necessary. For example, prolonged stress causes perturbation of the hypothalamic–pituitary axis and is a strong predictor of chronic pain, as noted in fibromyalgia. These harmful effects of stress damage other vital organs. For example, brain imaging studies demonstrated that people with increased stress had elevated activity in the amygdala and, five years later, had an increased risk of cardiovascular disease [20].

Post-traumatic stress disorder (PTSD) is a prime example of the impact that stress has on our brain and body as well as how it plays out in a bidirectional fashion. PTSD is defined as prolonged mood and cognitive disturbances that follow a traumatic event. The lifetime prevalence of PTSD in the general population is about 5% to 10%. During the COVID-19 pandemic, the prevalence of PTSD increased to 20% to 30% in the general population as did the prevalence of depression and anxiety disorders [21]. Widely divergent prevalence numbers have been reported in SARS-CoV-2 infection survivors, but persistent anxiety, stress, and PTSD, are described in 25% to 50% of patients [21, 22]. In the general population and long COVID, the increase in PTSD, depression, anxiety, cognitive disturbances, and sleep disturbances go hand-in-hand, mirroring their common biopsychological pathways.

Alterations in brain structure and function are well-documented in PTSD. Stress levels correlate with changes in the organization of various brain networks [23]. Lower volumes of brain regions, including the amygdala and the hippocampus, are present in subjects with PTSD, compared to controls, and correlate with the severity of mood and cognitive disturbances [24]. Functional MRI studies in PTSD have also demonstrated differences in regional brain connectivity compared to controls [25].

The damage from prolonged stress correlates with ANS dysfunction, termed dysautonomia. There are various forms of ANS dysfunction, each mechanistically related to inadequate blood vessel constriction when standing. This results in blood pooling in the lower part of the body and a reduced return to the heart, which triggers a compensatory increased heart rate and the release of hormones such as adrenaline.

For the body to function when there is an acute change in posture, when blood temporarily pools in the lower extremities, and there is a transient decrease in blood to the brain, the ANS receives information rapidly and must adjust. This requires blood pressure receptors in both the aorta, the main artery exiting the heart, and the carotid arteries, to make adjustments that bring blood to the brain. These are called baroreceptors. They must properly sense the drop in blood pressure then signal the sympathetic ganglia to speed up the heart. They also stimulate leg muscles to squeeze more blood out of the legs and ensure that there is sufficient water in the blood to immediately perfuse the brain and other vital organs. The baroreceptors also signal the adrenal gland to secrete epinephrine to help accomplish all these tasks. This is all done without our brains' conscious awareness, which only comes into play if we feel lightheaded or dizzy and are quick enough to lie or sit down to avoid passing out.

If the baroreceptors are damaged during the COVID-19 illness, similar to odor and taste receptors, they may not be able to detect the low-pressure signal. Even if they are able to receive the signal, the circuits that they try to activate must all be functioning normally to mount the proper response, and all this is occurring without our awareness, otherwise things could become dangerous.

Imagine a fire in a house where the smoke detectors do not work. Perhaps they detect the smoke, but the alarm does not sound. Or the alarm sounds but is not transmitted to the nearby fire station. Or the alarm is transmitted, but the firemen cannot get their trucks to start, or they get to your house, but they cannot open the hydrants and there is no water in the line. This kind of sequence may be analogous to what our bodies are going through to keep the blood flowing to our brains when we attempt to rapidly stand up from a lying position!

Patients with long COVID as well as CFS/ME and fibromyalgia have symptoms of ANS dysfunction, called dysautonomia. Tilt table or other postural testing has often reproduced the symptoms in long-COVID syndrome patients. In 180 patients with long COVID, 70% female and a mean age of 51 years, orthostatic intolerance during an active stand test was present in 14% [26]. The composite autonomic symptom scale 31 (COMPASS-31) was significantly elevated in more than one-quarter of the patients and was higher in female patients and those with neurologic symptoms. However, some studies did not find a correlation between ANS dysfunction and other long-COVID symptoms, such as fatigue [27].

The abnormality of the autonomic system most consistently linked to long COVID, CFS/ME, and fibromyalgia, is the postural orthostatic tachycardia syndrome (POTS). POTS is defined as a sustained elevation of the heart rate of at least 30 beats per minute when going from a supine to a standing position in the absence of a fall in blood pressure. It is associated with CNS symptoms, including fatigue, headaches, palpitations, dizziness, cognitive and sleep disturbances, and has many similarities to long-COVID syndrome. Initially described by Da Costa as "the soldier's heart" in USA civil war soldiers, like CFS/ME and fibromyalgia, it is considered a syndrome and an invisible illness [28]." POTS is quite common in the general population, mostly in young females.

Some individuals with POTS have a distal, small fiber sensory neuropathy postulated to be part of the POTS disorder [29, 30]. This small fiber neuropathy occurs in approximately half of POTS patients, and there is some evidence that small fiber neuropathy in POTS may correlate with reduced cardiac innervation [30, 31]. POTS and small fiber neuropathy are also found in CFS/ME, fibromyalgia, and long-COVID syndrome, in keeping with brain homeostasis running amok in each of these disorders [30–33].

Summary

We have presented evidence from functional imaging studies and PET scans demonstrating cerebral metabolic changes and electrophysiological data that the functional networks in our brains are altered in long COVID. These findings are strikingly similar to those present in other unexplained illnesses without obvious organ damage, such as fibromyalgia and CFS/ME. In these conditions and long-COVID syndrome, the symptoms cluster together and tend to get worse or better together, suggesting a common pathway.

In Chapter 7, we will postulate that this pathway involves a complex immune attack that disrupts normal brain function. This neuroimmune scenario ties many of our themes together – not a definite answer to the problem of the long-COVID syndrome, but one of several mechanistic ideas about its cause. Finally, in Appendix 2, we will explore further research paths that might support what we propose. The ultimate proof of any medical theory is an ability to either prevent, treat, or cure the disease in question. Future studies will let us know if this hypothesis, and its hint at novel therapies, will significantly aid those suffering from the mystery of the long-COVID syndrome.

References

1 National Institute for Health and Care Research (NIHR). (2020). NIHR Themed Review – Living with Covid-19. https://doi.org/10.3310/themedreview_41169

2 Ando, J., Milde, C., Diers, M. et al. (2020). Assessment of cortical reorganization and preserved function in phantom limb pain: a methodological perspective. *Sci. Rep.* 10: 11504.

3 Sluka, K.A. and Clauw, D.J. (2016). Neurobiology of fibromyalgia and chronic widespread pain. *Neuroscience* 338: 114.

4 Gracely, R., Petzke, R., Wolf, J.M., and Clauw, D.J. (2002). Functional magnetic resonance imaging evidence of augmented pain processing in fibromyalgia. *Arthritis Rheumatism* 46: 5.
5 Maksoud, R., du Preez, S., Eaton-Fitch, N. et al. (2021). A systematic review of neurological impairments in myalgic encephalomyelitis/chronic fatigue syndrome using neuroimaging techniques. *PLoS One* 15: e0232475.
6 Hsiao, F.-J., Chen, W.-T., Ko, Y.-C. et al. (2020). Neuromagnetic amygdala response to pain-related fear as a brain signature of fibromyalgia. *Pain Ther.* 9: 765.
7 Holmes, A.J. and Pizzagalli, D.A. (2008). Response conflict and fronto-cingulate dysfunction in unmedicated participants with major depression. *Neuropsychologia* 46: 2904–2913.
8 Douaud, G., Lee, S., and Alfaro-Almagro, F. (2022). Sars-CoV-2 is associated with changes in brain structure in UK Biobank. *Nature* 604: 697–707. https://doi.org/10.1038/s41586-022-04569-5.
9 Du, Y.-Y., Zhao, W., Zhou, X.-L. et al. (2022). Survivors of COVID-19 exhibit altered amplitudes of low frequency fluctuation in the brain: a resting-state functional magnetic resonance imaging study at 1-year follow-up. *Neural Regen. Res.* 17: 9.
10 Qin, Y., Wu, J., Chen, T. et al. (2021). Long-term microstructure and cerebral blood flow changes in patients recovered from COVID-19 without neurological manifestations. *J. Clin. Invest.* 131.
11 Benedetti, F., Palladini, M., Paolini, M. et al. (2021). Brain correlates of depression, post-traumatic distress, and inflammatory biomarkers in COVID-19 survivors: A multinational magnetic resonance imaging study. *Brain Behav. Immun. Health* 18: 100387.
12 Mavrikaki, M., Lee, J.D., Solomon, I.H., and Slack, F.J. (2021). Severe COVID-19 induces molecular signatures of aging in the human brain. *medRxiv* https://doi.org/10.1101/2021.11.24.21266779.
13 Sollini, M., Morbelli, S., Ciccarelli, M. et al. (2021). Long COVID hallmarks on [18F]FDG-PET/CT: a case-control study. *Eur. J. Nucl. Med. Mol. Imaging* 48: 3187–3197.
14 Guedj, E., Campion, J.Y., Dudouet, P. et al. (2021). 18F-FDG brain PET hypometabolism in patients with long COVID. *Eur. J. Nucl. Med. Mol. Imaging* 48: 2823–2833.
15 Toniolo, S., Scarioni, M., Di Lorenzo, F. et al. (2021). Dementia and COVID-19, a Bidirectional Liaison: Risk Factors, Biomarkers, and Optimal Health Care. *J. Alzheimers Dis.* 82: 883–898.
16 Hosp, J., Dressing, A., Blazhenets, G. et al. (2021). Cognitive impairment and altered cerebral glucose metabolism in the subacute stage of COVID 19. *Brain* 7: 144.
17 Versace, V., Sebastianelli, L., Ferrazzoli, D. et al. (2021). Intracortical GABAergic dysfunction in patients with fatigue and dysexecutive syndrome after COVID-19. *Clin. Neurophysiol.* 132: 1138–1143.

18 Tian, T., Wu, J., Chen, T. et al. (2022). Long-term follow-up of dynamic brain changes in patients recovered from COVID-19 without neurological manifestations. *JCI Insight* 22: 7.

19 Abdallah, C.G. (2021). Brain networks associated with COVID-19 risk: Data from 3662 participants. *Chron. Stress* 5: 1.

20 Tawakol, A., Ishai, A., Takx, R. et al. (2017). Relation between resting amygdala activity and cardiovascular events: a longitudinal and cohort study. *Lancet* 389: 834.

21 Nochaiwong, S., Ruengorn, C., Thavorn, K. et al. (2021). Global prevalence of mental health issues among the general population during the coronavirus disease-2019 pandemic: a systematic review and meta-analysis. *Sci. Rep.* 11: 10173.

22 Jafri, M.R., Zaheer, A., Fatima, S. et al. (2022). Mental health status of COVID-19 survivors: a cross sectional study. *Virol. J.* 6: 19.

23 Suo, X., Lei, D., Li, W. et al. (2020). Individualized prediction of PTSD symptom severity in trauma survivors from whole-brain resting state functional connectivity. *Front. Behav. Neurosci.* 14: 563152.

24 Basavaraju, R., France, J., Maas, B. et al. (2021). Right parahippocampal volume deficit in an older population with posttraumatic stress disorder. *J. Psychiatr. Res.* 137: 368.

25 Nisar, S., Bhat, A., Hashem, S. et al. (2020). Genetic and neuroimaging approaches to understanding post-traumatic stress disorder. *Int. J. Mol. Sci.* 21: 4503.

26 Stella, A., Furlanis, G., Frezza, N. et al. (2022). Autonomic dysfunction in post-COVID patients with and without neurological symptoms: a prospective multidomain observational study. *J. Neurol.* 269: 587.

27 Townsend, L., Moloney, D., Finucane, C. et al. (2021). Fatigue following COVID-19 infection is not associated with autonomic dysfunction. *PLoS One* 16: e0247280.

28 Kavi, L. (2022). Postural *tachycardia syndrome and long COVID. Br. J. Gen. Prac.* 72: 714.

29 Billig, S., Schauermann, J., Rolke, R. et al. (2020). Quantitative sensory testing predicts histological small fiber neuropathy in postural tachycardia syndrome. *Neurol. Clin. Pract.* 10 (5): 428–434.

30 Gibbons, C., Bonyhay, I., Benson, A. et al. (2013). Structural and Functional Small Fiber Abnormalities in the Neuropathic Postural Tachycardia Syndrome. *PLoS One* 8: 12, e84716.

31 Haensch, C., Tosch, M., Katona, I. et al. (2014). Small fiber neuropathy with cardiac denervation in postural tachycardia syndrome. *Muscle Nerve* 50: 6.

32 Novak, P., Mukerji, S., Alabasi, H. et al. (2021). Multisystem involvement in post-acute sequelae of COVID-19 (PASC). *Ann. Neurol.* 91: 3.

33 Blitshteyn, S. and Whitelaw, S. (2021). Postural orthostatic tachycardia syndrome (POTS) and other autonomic disorders after COVID-19 infection: a case series of 20 patients. *Immunol. Res.* 1: 6.

7

Neuroimmune Dysfunction

Is Long COVID an Autoimmune Disease?

Many investigators have suggested that long COVID is an autoimmune disease [1]. Classic autoimmune diseases, such as rheumatoid arthritis, systemic lupus erythematosus (SLE; lupus), and multiple sclerosis (MS), are characterized by a misdirected and overzealous immune reactivity. High levels of autoantibodies are almost always present in the blood, and immune deposits with intense inflammation are typically observed at the damaged organs.

This is what happens during an acute SARS-CoV-2 infection. There is a powerful immune response, commonly dubbed the "cytokine storm", with very high levels of cytokines and systemic inflammation reflected by high levels of C-reactive protein (CRP) and D-dimer. This is present even in non-hospitalized patients, since patients with mild or moderate COVID-19 have a marked increase in autoantibody reactions compared to uninfected individuals, and a high prevalence of autoantibodies against cytokines, chemokines, and other immune modulators [2]. This immune process is responsible for the subsequent organ damage. Dr. Steven Deeks, an infectious-disease researcher at the University of California, San Francisco said "SARS-CoV-2 is like a nuclear bomb in terms of the immune system. It just blows everything up" [3].

The term cytokine storm is used widely in both scientific and lay articles about the severe consequences of COVID-19. It is such a common term that one might imagine everyone interested in the COVID-19 disease and the long-COVID syndrome would know exactly what cytokines are and how they affect the immune system and the brain. Unfortunately, understanding cytokines and the cytokine storm is a challenge for both physicians and lay people alike.

Cytokines are small proteins secreted mostly (but not exclusively) by immune blood cells. There are more than 100 different cytokines. Cytokines are often

Unravelling Long COVID, First Edition. Don Goldenberg and Marc Dichter.

secreted by immune cells when the body is stressed or injured. Most often they act on other immune cells, but they are capable of interacting with possibly every cell in the body. Some cytokines have effects that are consistently focused on one process, such as inflammation. Sometimes a given cytokine can be pro-inflammatory or anti-inflammatory, depending on the tissue and local environment. Cytokines often act on other cells to secrete additional cytokines in complex regulatory networks. Furthermore, the same cytokine may act differently according to the organism's sex [1, 4].

This makes it difficult to unequivocally determine the role or level of any given cytokine in blood or cerebrospinal fluid (CSF) unless a specific target is identified and a functional test exists. Cytokines may also stimulate or suppress cell division in different organs, including the brain. Another class of cytokines, known as chemokines, function to direct immune cells toward areas of inflammation or damage.

In a cytokine storm, the massive outpouring of inflammatory cytokines produces more organ damage than the viral infection itself. Cytokine storms must be actively treated by drugs designed to reduce inflammation, such as corticosteroids, rather than antiviral drugs that would no longer be useful. Cytokines enhance inflammatory activity in the lungs, heart, gastrointestinal system, kidneys, and other body organs. The effects of a cytokine storm on the brain are not well understood but may be controlled by the cytokines' limited ability to cross the blood brain barrier.

A less severe activation of cytokines occurs in patients with mild, or even, asymptomatic COVID-19 infections, and there are increased circulating cytokines and autoantibodies compared to people without evidence of having had COVID-19 [5].

Another indication that long COVID has an autoimmune basis is the fact that it is much more common in younger women. So are the classic autoimmune diseases, including rheumatoid arthritis, SLE, and MS. The immune reactivity of females from many species, including humans, is more advanced than their male counterparts [6]. The increased mortality from an acute SARS-CoV-2 infection in male patients may be related to their less effective acute immune response. However, a heightened, prolonged immune reaction in female patients may account for the increased prevalence of long-COVID syndrome in younger women. There does appear to be sex-related differences in the autoantibody response following mild or asymptomatic SARS-CoV-2 infection that may emerge months after the infection and account for the female predominance in long COVID [7].

Various infectious agents may initiate an autoimmune disease, such as streptococci that causes rheumatic fever or hepatitis virus that causes polyarthritis. MS is an autoimmune disease long-been considered to be triggered by an infection. The prime candidate has been the Epstein–Barr virus (EBV), but without definitive proof. EBV can cause multiple illness, including mononucleosis and several forms of lymphoid cancers. However, EBV is widespread so tying it directly to MS has been a challenge. A recent report analyzed more than 10 million recruits in the US military who were tested for prior EBV infection [8]. Among those, 955 were diagnosed with MS, almost all of whom had recently acquired EBV. The risk of MS in those who did not acquire EBV was 32-fold less. A biomarker for neuronal degeneration, the neurofilament light chain, was increased after EBV infection. The increased risk of MS and the biomarker were not associated with the acquisition of any other common chronic viral infections such as cytomegalovirus.

Following SARS-CoV-2 infection, the only clear evidence of a classic autoimmune disease has been the rare occurrence of multi-system inflammatory syndrome (MIS-C for children and MIS-A – for adults). This is clearly an autoimmune disease that follows on the heels of a SARS-CoV-2 infection and is more common in children. It is characterized by rash, myocarditis, and arthritis with abnormal laboratory tests and usually responds well to anti-inflammatory and immunosuppressive therapy. There is evidence of organ damage involving the heart, lungs, kidneys, and the brain. MIS resembles Kawasaki's disease, a multi-system, autoimmune condition that often follows a viral infection.

We discussed Guillain-Barre syndrome (GBS) in Chapter 3, where autoantibodies, presumed to be directed at viruses, can cross react with myelin on peripheral nerves to cause severe neuropathy. GBS may be associated with SARS-CoV-2 and, if that association is proven, it would be another example of an autoimmune disease triggered by SARS-CoV-2. Other examples of autoantibodies to neuronal cells have been known about for many years. Patients with various forms of cancer have developed peculiar neurological disorders that were discovered to be caused by autoantibodies directed against the tumor cells that also appear to, by mistake, attack neurons in the brain or peripheral nervous system.

An example of a central nervous system (CNS) autoimmune disorder that is relatively well understood is *N*-methyl-d-aspartate (NMDA)-receptor encephalitis. Patients with this disorder may present with a psychosis, proceed to develop severe seizures, and then lapse into coma. Patient MRIs show a nasty inflammatory lesion in the limbic lobes. A small group of women with this syndrome were noted to have a benign ovarian tumor composed of many body-cell types, including brain cells. Initially, physicians did not see any relationship with the tumors. Subsequently, it was discovered that the patients had antibodies in their blood that reacted with brain cells [9]. When the tumors were removed, the antibodies declined and patients, some of whom were in a coma for years, recovered. The antibodies responsible for the disease were directed at the NMDA receptor on the surface of many neurons, especially in the hippocampus and surrounding areas [9]. The NMDA receptor is an important component of excitatory synapses throughout the brain and appears to play a major role in memory. The antibodies did not kill or severely injure the affected neurons but caused the NMDA receptors to be internalized, thereby significantly disrupting these cells' brain circuits and apparently causing the disease. Removing the benign tumors caused the antibodies to disappear, which allowed the NMDA receptors to return to the membrane surfaces and the circuits to function normally.

As this autoimmune disease was recognized in more patients, many were found without tumors or a clear inciting cause, but the pathobiology was similar and they responded to appropriate anti-inflammatory treatments. Additional studies revealed antibodies against other nerve cell components that produced similar, but not identical results. There is evidence that this autoimmune disorder can be triggered by a viral infection.

We believe there is overwhelming evidence that acute SARS-CoV-2 infections and long-COVID disease have an autoimmune component that is important in progressive organ damage. The question remains whether there is a brain-specific immune pathway that helps explain long-COVID syndrome in the absence of organ damage.

Evidence that Long-COVID Syndrome is a Neuroimmune Disorder

We believe that rather than a classic, autoimmune disease, long-COVID syndrome is best thought of as a neuroimmune disorder, without evidence of a *systemic*, autoimmune disease. In contrast to a disease like SLE, there is no defined organ damage, either in the body or the brain (Table 7.1). This neuroimmune attack is localized to the CNS and leads to alterations in normal brain pathways. CNS dysfunction, rather than damage, is the result.

Table 7.1 Characteristics of a systemic autoimmune disease versus neuroimmune disorder.

	Systemic autoimmune disease	**Neuroimmune disorder**
Organ damage	Yes, brain and systemic	Brain dysfunction
Inflammatory biomarkers	Yes (CRP, D-dimer)	No – but cytokines, autoantibodies in CSF
Immune markers	Yes	Yes – different from those in blood
Activated microglia	Yes	Yes – may be regional
Examples	SLE, MS	CFS/ME, fibromyalgia

Since there is no organ inflammation, the immune/inflammatory disease markers universally present in SLE and MS would not be present in long-COVID syndrome. Because there is no organ damage, tests such as imaging the brain or other organs would be unrevealing. Conditions such as CFS/ME, fibromyalgia, and depression could also be considered neuroimmune disorders.

Neuroimmunology is dedicated to understanding the interaction between the immune and nervous systems, demonstrated clearly in neurologic diseases such as MS and myasthenia gravis. There is great interest in applying neuroimmunology to a variety of neurologic and psychiatric disorders, sometimes called psychoneuro-immunology, which involves appreciating bidirectional mind-body interactions.

A brief review of neuroimmune mechanisms helps explain why we believe this field may provide much needed answers to unravelling long-COVID syndrome. In a neuroimmune model, SARS-CoV-2 activates a localized immune response that manifests primarily in the CNS. Dr. Robert Dantzer, Professor of Integrative Immunology and Behavior at the University of Illinois, said that a brain-immune response behaves as, "... the unsuspected conductor of the ensemble of neuronal circuits and neurotransmitters that organize physiological and pathological behaviour" [10]. This results in endocrine, autonomic, and behavioral changes triggered by chemical mediators that include cytokines produced by brain cells, such as microglia. Microglia are brain-immune cells [11]. These cells that react to infection, brain damage, and other forms of brain stress, including psychological stress [11]. They rapidly proliferate, change their molecular composition, and help regulate brain homeostasis. The microglia are activated by cytokines and secrete other cytokines that act on other brain cells and cells of the immune system. Activated microglia can be readily recognized with pathological examinations of the brain in humans, and more easily in experimental animals where the activation of the microglia can be examined in detail as it occurs. Activated microglia can be localized in living humans by a sophisticated positron-emission tomography (PET) scan technique using a radioactively labeled compound that binds specifically to the activated microglia [12, 13].

In long-COVID syndrome, there is no dramatic *systemic*, immune inflammatory response accounting for the largely normal inflammatory and autoimmune markers and absence of detectable organ damage. We believe that a difficult to detect, relatively brain-specific, immune reaction may help to explain long COVID. This involves microglial activation in specific brain regions as well as specific cytokine secretion from those areas, resulting in profound effects on brain function and symptoms of long-COVID syndrome (Figure 7.1) [14].

The brain's innate microglia as well as microglia that arise from circulating macrophages in the brain respond to distress signals from infection or from any major stress [4]. There is no direct invasion of immune cells, so the evidence of

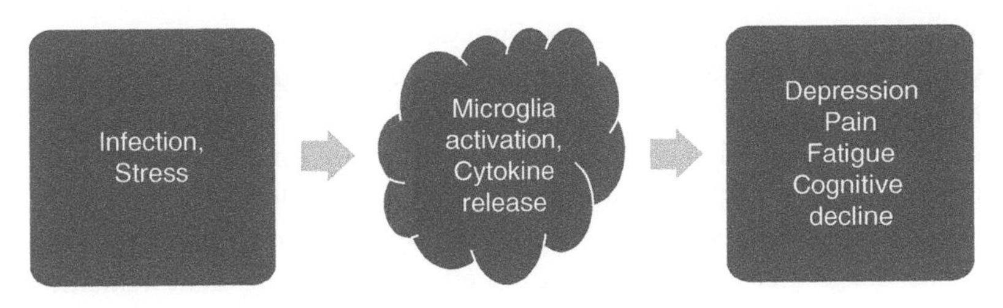

Figure 7.1 Model of localized neurohormone reactivity in long COVID. *Source:* Based on [4, 14].

brain damage is not evident. This brain-specific immune reaction is consistent with lack of systemic inflammation, either in blood tests or organ damage. Microglia secrete pro-inflammatory cytokines that have very potent effects on brain cells and can cause both dysfunction and damage. Such microglial activation and cytokine release would result in depression, pain, fatigue, and cognitive disturbances. Chronic stress alters the size, shape, and density of microglia in specific brain regions that may lead to mood and cognitive disturbances (Figure 7.1) [4, 14].

Cytokines and the Brain

Cytokines are important in patients that experience neuropsychiatric symptoms during, and long after chemotherapy, known as chemo brain. The symptoms are similar to those reported in long COVID, including problems with memory, word retrieval, concentration, processing numbers, following instructions, multitasking, setting priorities, and is associated with severe fatigue [15]. A cancer patient described her symptoms to Jane Brody in *The New York Times*, "Inability to focus on anything with any complexity or depth, inability to retain information, especially names, difficulty retrieving words and substituting wrong words ('chicken' for 'kitchen'), difficulty analyzing anything other than simple questions and inability to follow instructions when cooking or knitting" [16].

Difficulties objectively defining and measuring the effects of chemotherapy on the CNS result in vast differences when estimating the percentage of cancer survivors with chemo brain, which range from 17% to 75% [17]. In general, there is some recovery that occurs after chemotherapy stops, but deficits in brain function can be detected up to 10 years after treatment, suggesting that they are permanent in some cancer survivors. Structural MRI studies in chemo brain reveal decreased gray matter density in several brain regions, including the frontal and temporal cortices, the cerebellum, and the right thalamus immediately after chemotherapy, with only partial recovery a year later [18] Functional magnetic resonance imaging (fMRI) studies also found decreased activation during cognitive tasks in similar regions [19].

It is likely that the underlying mechanisms of chemo brain are different from those underlying long-COVID syndrome, but some of the down-stream consequences may be analogous. Chemotherapy for systemic malignancies often employs toxic drugs that can have off-target effects, including neurotoxicities. These could, in turn, initiate brain-specific dysfunctions mediated by microglia and cytokines that produce network disruptions like those noted in the long-COVID patients. Immunotherapy may also trigger an associated cytokine storm similar to that in COVID-19. The convergence of these symptom patterns among these separate disorders supports their common pathway and potential relationship to symptoms such as depression, chronic pain, sleep disorder, brain fog, and chronic fatigue.

In addition to their effects as modulators of the immune system, cytokines can also impact the normal functions of microglia, astrocytes, and neurons in the brain [20]. This may contribute to the disrupted brain networks seen in long-COVID syndrome and conditions such as CFS/ME [21] and fibromyalgia [22]. Cytokines are not easily able to enter the brain because of the blood brain barrier. However, there are some transport systems that carry some cytokines into the CSF surrounding the brain. Furthermore, there are cells from the lymphoid immune system that exist under the membranes that surround the brain and they can secrete cytokines directly into the CSF. In the brain, endogenous microglia and astrocytes secrete cytokines and these cells are also activated by cytokines secreted by immune cells in the blood and in the fluid around the brain.

Cytokines can alter brain function by interfering with important synaptic communication, including reducing and possibly increasing normal connectivity in brain networks under different circumstances, even in the absence of inflammation or cell damage. Some cytokines, for example, interleukin-4 (IL-4), are associated with cognition [23] and appear to enhance cognitive function [23]. Knocking out IL-4 in mice produces animals with poor memories in various tests. Restoring the IL-4 in these mice enhances memory. Interestingly, levels of IL-4 are also reduced in patients with dementia [23]. Whether this is a part of the pathophysiology underlying dementing, neurodegenerative disorders in humans is unclear. In any case, if IL-4 is low in patients with long-COVID syndrome, it may provide a clue as to the brain fog so often described in these patients. It could also point toward a specific treatment for this condition.

IL-6 is another cytokine dysregulated in both acute COVID-19 and patients with the long-COVID syndrome [24]. IL-6 is generally considered a pro-inflammatory cytokine. In severe, acute SARS-CoV-2 infection, IL-6 was thought to be heavily involved in the inflammatory damage, and at least two clinical trials were conducted to determine if monoclonal antibodies directed at blocking the effects of IL-6 would be a useful therapy for very ill patients [24]. One showed that this treatment worked while the other did not. However, the trial conditions were significantly different, especially with regard to the timing and administration of corticosteroids during the cytokine storm, suggesting that this approach might be useful in conjunction with other anti-inflammatory strategies. IL-6 is elevated in patients with long-COVID syndrome, along with other cytokines, including IL-1β and tumor necrosis factor-α (TNFα). How these cytokines may contribute to long-COVID syndrome symptoms is not yet known. However, the availability of therapeutic agents to neutralize each of these cytokines may lead to novel long-COVID therapies.

Another cytokine/chemokine elevated in patients after recovering from mild COVID-19 without known CNS invasion of the virus, and in mice infected with SARS-CoV-2, is CCL11 [1]. This cytokine usually attracts eosinophils to specific locations, and it is implicated in allergic reactions. In the brain, CCL11 is associated with impaired neurogenesis in the hippocampus and reduced cognitive function. In one study, elevated CCL11 was found in 48 individuals experiencing cognitive dysfunction after COVID-19 compared to levels in 15 patients who did not have cognitive dysfunction. In the mouse model of mild SARS-CoV-2 infection, a reduction in hippocampal neurogenesis compared to controls was noted, as well as microglial activation. Each of these effects lasted more than seven weeks post-infection. Taking all these data together, investigators concluded that a "profound multi-cellular dysregulation in the brain was caused by even mild respiratory SARS-CoV-2 infections" [1]. These researchers have also reported a distinct pattern of microglial activation in white matter that is associated with cancer therapy-related cognitive impairment, a syndrome associated with increased inflammatory cytokine IL6 and chemokines.

Autoantibodies in Long-COVID Syndrome

Important evidence for autoimmunity in patients with long-COVID relates to the presence of antibodies in the blood or CSF that are directed against normal brain tissue [1, 2, 5]. Other examples of autoantibodies to neuronal cells in non-COVID diseases have been known for years. Distinct from chemo brain, patients with various forms of cancer have developed peculiar neurological disorders discovered to be caused by autoantibodies directed against tumor cells but appear to also, by mistake, attack neurons in the brain or peripheral nervous system. These disorders, called paraneoplastic disease, may cause major inflammatory reactions in the affected brain regions. Some are associated with significant cell losses in restricted regions (e.g. cerebellar Purkinje cells in paraneoplastic cerebellar degeneration) but not necessarily with easily identified inflammation [25]. In most cases, these disorders do not respond to classical anti-inflammatory drugs. However, the autoimmune disease may disappear if the associated tumors can be successfully removed. Occasionally, there are cases where after an apparent cure of the original cancer, the cancer recurs years later along with the paraneoplastic neurological disease.

Song et al., found anti-virus antibodies in the CSF of long-COVID patients as well as increased T-cell activation, although they did not find viral RNA in CSF [5]. They then investigated serum versus CSF levels of cytokines and antibodies in transgenic mice with human ACE-2 receptors in either the lung or brain, or in both. The mice with SARS-CoV-2 infections restricted to the lungs had antiviral antibodies in the blood, but not in the CSF. Mice with SARS-CoV-2 infections restricted to the brain had antiviral antibodies in the CSF, but not in the blood. While they expected to detect viral RNA in the CSF of mice with brain infections, they did not, despite the presence of antiviral antibodies that appeared to be produced by immune cells in the brain. These data were surprising and clearly demonstrated a two-compartment system for the immune response against infections in the body and the brain. The source of viral antigens in the brain compartment remains unclear. They used monoclonal antibodies to neuronal tissue derived from cloned, human CSF B cells to stain neurons in mouse brain sections, including in the cortex, CA3 region of the hippocampus, olfactory bulb, thalamus, cerebellum and brain stem. Staining was also seen in cerebral blood vessels [5]. These studies indicate that brain-specific immune reactions may be important in understanding long COVID. However, more studies of immune reactivity in both the serum and CSF will be essential.

In one study, COVID-19 patients had unique autoantibodies against HCRTR2, an orexin receptor enriched in hypothalmic neurons [2]. Orexin is released by neurons in the hypothalamus to promote normal wakefulness, so blocking orexin's actions at its receptor might interfere with sleep. In fact, potential new drugs to treat insomnia have been developed to target the orexin receptor. COVID-19 patients with these orexin blocking autoantibodies had *decreased* levels of consciousness during their illnesses compared to comparably ill patients without these specific autoantibodies. Based on these observations one can only imagine the scope of other, as yet unidentified, cytokines and autoantibodies that may develop in COVID-19 patients and their direct effects on brain function and network reorganization.

Other autoantibodies in patients with persistent long-COVID symptoms include one report in which all 29 symptomatic, post-COVID-19 patients had autoantibodies directed against a group of G-protein coupled receptors involved in neuron-to-neuron signal transmission and network regulation [26]. Both the muscarinic acetyl choline receptor and the beta-adrenergic receptor, which are critically involved in autonomic responses, were antibody targets. Some of these antibodies acted as agonists to directly activate the receptors. Autoantibodies that alter neuron-receptor function could clearly disturb carefully regulated autonomic functions and induce several long-COVID symptoms such as disturbances in heart rate, blood pressure control, breathing, and the development of POTS.

Can Vaccines also Cause Autoimmune Reactions?

Vaccines against SARS-CoV2 can cause transient flu-like symptoms that usually last for hours or a few days, including a low-grade fever, loss of appetite, lack of energy, headaches, and other symptoms. Immune responses themselves can mimic viral illnesses and some of the symptoms of the long-COVID syndrome, albeit transiently. It is likely that some of the disease manifestations that we commonly ascribe to a cold virus or the flu are caused by our body's immune responses to protect ourselves from the virus.

There are a small number of patients who reported more substantial and disabling symptoms after vaccination with both the mRNA- and attenuated virus-based vaccines, which suggests that SARS-CoV-2 vaccines may potentially cause some, transient long-COVID symptoms [27]. Vaccines can also, rarely, cause systemic immune reactions such as an inflammatory myocarditis, which can usually be treated with anti-inflammatory drugs. Vaccinations are also associated with blood clots, but the mechanism is unknown. These cases are rare and sufficiently treatable that the risk–benefit analysis of a vaccine against a potentially serious, and possibly fatal, disease is such that most physicians and public health officials continue to make strong appeals for the general public to be vaccinated against COVID-19.

Immune Tests for Long COVID

There is great interest in using immune tests to predict whether a patient will develop long COVID during, or shortly after, the initial SARS-CoV-2 infection [28]. A group of investigators extensively studied the immune statuses of patients during the initial SARS-CoV-2 infection and two to three months later [29]. Most of these patients were hospitalized, and one-third had symptoms consistent with long COVID at three months. The most influential factor to predict long COVID were autoantibodies, which were detected in two-thirds of the long-COVID patients. Other correlations with long COVID included evidence for the reactivation of Epstein–Barr virus and alterations in the hypothalamic-pituitary-adrenal (HPA) axis.

Such studies, done two or three months after infection, need to be carried out much longer, and with similar tests on the CSF, to determine whether these immune signatures for long COVID persist. The results may simply reflect the increased immune reactivity during the acute infection. Furthermore, most of the patients in this study were hospitalized, making it difficult to generalize these results to the larger population.

In a recent report [30], there was no correlation between developing long COVID in SARS-Cov-2 infected patients (most of whom were not hospitalized) with plasma levels of multiple immune markers including CRP, D-dimer; biomarkers of cardiac injury like troponin 1; or brain injury, including neurofilament light chain. Nor was there a correlation with standard autoimmune tests, such as antinuclear or anticardiolipin antibodies and rheumatoid factor, or selected immune/inflammatory markers, including plasma levels of macrophage inflammatory protein-1β, IFN-γ, TNF-α, programmed cell death ligand-1, IFN-γ–induced protein 10, IL-2 receptor α, IL-1β, IL-6, IL-8, RANTES (regulated on activation, normal T-cell expressed and secreted), and CD40.

There is overwhelming evidence that long-COVID disease has an autoimmune component important for understanding progressive organ dysfunction in the absence of detectable damage. The question remains whether there is a brain-specific immune component that helps explain long-COVID syndrome and the absence of organ damage. If the immune dysregulation during long-COVID syndrome is limited to the CNS, the characteristic biomarkers of systemic autoimmune disease would not be present (Table 7.1). There would be no detectable brain damage, as is usually seen with imaging studies of SLE with brain dysfunction. However, neuroimmune-induced brain dysfunction could be responsible for the persistent exhaustion, pain, mood, cognitive, and sleep disturbances of long-COVID syndrome as well as autonomic nervous system (ANS) alterations such as POTS. SARS-CoV-2 infected, or otherwise activated, microglia could release cytokines initiating this process, independent of any systemic immune disorder.

The evidence to correlate the unique characteristics of CSF cytokines and autoantibodies with the symptoms of long-COVID syndrome and their intensity is beginning to present itself, but a lot remains to be learned. For example, the data from Wang et al. demonstrating reduced alertness in patients with antibodies against the orexin receptor shows how the direct effects of one autoantibody in COVID-19 patients may alter disease severity by affecting a prominent neurological pathway [2]. Cytokines and antibodies against cytokines likely have several avenues through which to alter brain function that need to be elucidated. Excess autoantibodies and cytokine presence occurs in very ill COVID-19 patients as well as mildly ill or asymptomatic individuals, both in patient blood and CSF. Research from multiple laboratories is revealing connections between long-COVID syndrome symptoms and the effects of autoantibodies or cytokines on the brain.

Although much of our discussion in this chapter is focused on patients with long-COVID syndrome whose symptoms cannot be attributed to the organ damage occurring during the acute COVID-19 disease, it is also important to recognize that patients with long-COVID disease and visible brain damage may also develop additional neuroimmune factors, including activated CNS-specific cytokines and autoantibodies similar to those of patients with the long-COVID syndrome. These may then produce unexpected longer term medical issues during recovery.

Neuroimmune Studies in CFS/ME, Fibromyalgia, and Depression

Inflammatory cytokines are implicated in chronic pain by their effect on the reorganization of pain signals from the peripheral sensory source to the perception of pain in the cerebral cortex [21, 31]. Increased pro-inflammatory cytokines and glial activation occurs in patients with CFS/ME and fibromyalgia [21, 22], who also have elevated levels of TGF-β and pro-inflammatory cytokines, including CCL11, CXCL10, IFN-γ, and IL-5 [32, 33]. In CFS/ME, excess fatigue correlates with increased cytokine production and activated microglia [33]. Alterations in B- and T-cell subsets and increased levels of autoantibodies, including antinuclear and anticardiolipin antibodies characteristic of SLE, also occur in CFS/ME [33].

Antibodies against two neurotransmitter receptors, muscarinic M1 acetylcholine receptor and beta-adrenergic receptor, were reported in patients with long-COVID syndrome and a subset of CFS/ME patients [26, 34]. Increased CSF levels of chemokines involved in neuron-glial communication were present in fibromyalgia patients [31]. Inflammatory cytokines are also implicated in depression [35, 36]. Individuals with treatment-resistant depression have activated microglia in localized brain regions and increased cytokines in their CSF. IFN-α, IL-1, and IL-2 appear to promote depression. Other cytokines, such as TNF-α and IFN-γ appear to have antidepressant effects. As mentioned in Chapter 6, there is abundant evidence that the hippocampus and other brain structures appear smaller in individuals with treatment-resistant depression, and this apparent shrinkage is at least partially reversed if the depression is effectively treated.

Is a Unifying Theory of How Long-COVID Syndrome Develops Possible?

At this point, we will propose a unifying hypothesis that brings together much of the evidence presented in Chapters 6 and 7 to posit an explanation of the many symptoms in long-COVID syndrome patients who have minimal or no evidence for organ damage in the brain, lungs, or heart to otherwise explain their symptoms. This is not the only possible hypothesis for long-COVID syndrome, but it does present some novel ideas and suggest novel treatments.

There is a small cluster of nerve cells just under the pineal gland in the human brain called the habenula. This part of the brain is present in nearly all vertebrate species, including mice, fish, and lampreys. Its faithful preservation over millions of years of evolution signifies that it is a phylogenetically ancient and important brain structure. Neurons in the habenula are the only ones that can produce and release the cytokine IL-18. Ependymal cells that line the lateral and medial ventricles of the brain also produce IL-18, as do activated microglia. IL-18 is an inflammatory cytokine implicated in multiple systemic autoimmune diseases such as diabetes and rheumatoid arthritis as well as at least one brain disease – MS. In animal models, when acutely or chronically stressed, the habenula microglia are activated and produce IL-18, significantly increasing IL-18 in the habenula neurons but not the ependymal cells, indicating the habenula's role in the consequences of stress [37].

The habenula is overly activated by both physical and psychological stress. It is directly affected when the HPA is activated as part of the sympathetic nervous system's reaction to stress (the fight-or-flight reaction). The habenula's output is directly connected to multiple brain networks, including two prominent regions that (i) control reward signals via dopamine and (ii) regulate moods via serotonin. In both humans and mice alike, when the habenula is overactivated its axonal fibers inhibit, or down regulates, those areas. Connections from the habenula also influence sleep and circadian rhythms. The habenula also has connections to multiple ANS regulatory brain regions, the olfactory cortex, where the sense of smell is organized into recognized odors, and higher level pain pathways [37].

Habenula neurons also have ACE-2 receptors, the receptors that the SARS-CoV-2 virus uses to attach and infect cells. At the time of writing, there was no evidence regarding SARS-CoV-2 infection of habenula neurons. Habenula neurons also highly express NMDA receptors, which will become important to remember near the end of this section. When a mouse is stressed, the habenula becomes overactive and the mouse develops an adverse reaction to its environment [37]. It acts as if responding to danger, despite being in a neutral, harmless environment. This is a persistent reaction. Both mice and fish exhibit reduced spontaneous motor activity, as if they are freezing in place and cannot, or do not want to, move around. During habenula stimulation, mice also exhibit symptoms of depression similar to those commonly used in behavioral tests. Inhibiting overactivity in the habenula rapidly reverses all of these symptoms.

MRI techniques confirm that the human habenula is overactive after stress. If the human system acts similarly to the mouse brain, persistent overactivity of the habenula could produce a chronic fatigue whereby the muscles function but the brain creates a negative internal sense of ability or desire to exercise. This could also worsen symptoms if exercise is not carefully introduced, as is the case for many CFS patients. Such an individual would also feel depressed (a down-regulated serotonin mood modulating system), have difficulty sleeping, and maybe experience brain fog (confusion, decreased concentration, and decreased cognitive function). All of these are common symptoms of long-COVID syndrome. At the time of writing, there are no reports of fMRIs or any other imaging of the human habenula in patients with long-COVID syndrome. Similarly, there are no reports of habenula pathology in postmortem exams. This is not completely unexpected, as the habenula is a relatively small structure that is hard to image directly.

We present this hypothesis primarily to stimulate similar long-COVID research, not only involving the habenula, but also similar neural structures that may be involved in localized immune pathways. There may also be practical application of such theories. For example, ketamine, an NMDA receptor antagonist used for treatment-resistant depression in the general population, might reduce the neuronal activity of the habenula, considering the abundance of NMDA receptors on these neurons. There are scattered reports of using ketamine to treat fibromyalgia with variable results, but it may be reasonable to consider it in a trial for long-COVID management [38].

Summary

We believe that the long-COVID syndrome is best explained by alterations in normal brain homeostasis triggered by SARS-Cov-2 infection. There is no evidence for a persistent, ongoing infection, although it remains possible that some viral components are sequestered and difficult to find. We believe that, rather than a systemic, autoimmune disease, long-COVID syndrome is associated with immune activation that primarily affects the CNS. Brain-specific neuroimmune mechanisms result in structural and/or functional abnormalities in the CNS and the ANS. The increased levels of cytokines/chemokines in the blood, CSF, and brain tissue as well as the presence of activated microglia, autoantibodies against brain tissue, a lack of systemic inflammatory signals, absence of organ disease, all support this hypothesis.

Moreover, there is clear evidence that, viral infection aside, the additional significant stress that so many have experienced can produce physical changes in the brain and interrupt normal brain networks and physiology. The details of the network reorganizations discussed in Chapter 6 that make the brain run amok are not clearly understood, but the neuroimmune mechanisms provide a window for discovery. In Appendix B: Suggestions for Future Research Focused on the Cellular and Molecular Basis of the Long-COVID Syndrome we will integrate these mechanisms and pathways to discuss current and future evaluation and management of long COVID.

We postulate that the neuroimmune aberrations present in long-COVID patients causes the brain's dysfunctional state, as represented by the disabling symptoms of long-COVID syndrome, even in the absence of evidence for well-understood brain damage. We can then see the transition between a syndrome with an unexplained cause to a disease with a recognized cause and a potential path toward therapy. Hopefully, this same reasoning will be applicable to other disorders of unknown cause, such as fibromyalgia, CFS/ME and others that closely resemble the symptomatology of long-COVID syndrome.

References

1. Knight, J., Caricchio, R., Casanova, J.-L. et al. (2021). The intersection of COVID-19 and autoimmunity. *J. Clin. Invest.* 131 (24).
2. Wang, E.Y., Mao, T., Klein, J. et al. (2021). Diverse functional autoantibodies in patients with COVID-19. *Nature* 595: 283.
3. Marshall, M. (2021). The four most urgent questions about long COVID. *Nature* 594: 168–170.
4. Tynan, R., Beynon, S., Hinwood, M. et al. (2010). Chronic stress alters the density and morphology of microglia in a subset of stress-responsive brain regions. *Brain Behav. Immun.* 24: 1058.
5. Song, E., Bartley, C., Chow, R. et al. (2021). Divergent and self-reactive immune responses in the CNS of COVID-19 patients with neurological symptoms. *Cell Rep. Med.* 2.
6. Klein, S. and Flanagan, K. (2016). Sex differences in immune responses. *Nat. Rev. Immunol.* 16: 638.
7. Liu, Y., Ebinger, J.E., Mostafa, R. et al. (2021). Paradoxical sex-specific patterns of autoantibody response to SARS-CoV-2 infection. *J. Transl. Med.* 19: 524.
8. Bjornevik, K., Cortese, M., Healy, B. et al. (2022). Longitudinal analysis reveals high prevalence of Epstein-Barr virus associated with multiple sclerosis. *Science* 21 (375): 296–301.
9. Dalmau, J., Tuzun, E., Wu, H. et al. (2007). Paraneoplastic anti-N-methyl-D-aspartate receptor encephalitis associated with ovarian teratoma. *Ann. Neurol.* 61 (1): 25–36.
10. Dantzer, R., O'Connor, J., Freund, G. et al. (2008). From inflammation to sickness and depression: when the immune system subjugates the brain. *Nat. Rev. Neurosci.* 9: 46.
11. Matcovitch-Natan, O., Winter, D., Giladi, A. et al. (2016). Microglia development follows a stepwise program to regulate brain homeo-stasis. *Science* 353.
12. Beaino, W., Janssen, B., Vugts, D. et al. (2021). Towards PET imaging of the dynamic phenotypes of microglia. *Clin. Exp. Immunol.* 206: 282–300.

13 Kreisl, W., Kim, M.-J., Coughlin, J. et al. (2020). PET imaging of neuroinflammation in neurological disorders. *Lancet Neurol.* 19 (11): 940–950.

14 Tremblay, M.-E., Madore, C., Bordeleau, M. et al. (2020). Neuropathobiology of COVID-19: the role for Glia. *Front. Cell. Neurosci.* 14: 592214.

15 Nguyen L and Ehrlich B. (2020). Cellular mechanisms and treatments for chemo brain: insight from aging and neurodegenerative diseases. *EMBO Mol. Med.* 8: 12.

16 Brody J. (2009). The fog that follows chemotherapy. *The New York Times* (3 August).

17 Wefel, J., Kesler, S., Noll, K., and Schagen, S. (2015). Clinical characteristics, pathophysiology, and management of noncentral nervous system cancer-related cognitive impairment in adults. *CA Cancer J. Clin.* 65 (2): 123–138.

18 McDonald, B., Conroy, S., West, J. et al. (2012). Alterations in brain activation during working memory processing associated with breast cancer and treatment: a prospective functional magnetic resonance imaging study. *J. Clin. Oncol.* 30: 20.

19 Kesler, S., Rao, A., Blayney, D. et al. (2017). Predicting long-term cognitive outcome following breast cancer with pre-treatment resting state fMRI and random forest machine learning. *Front. Hum. Neurosci.* 15.

20 Zhang, J. and An, J. (2007). Cytokines, inflammation and pain. *Int. Anesthesiol. Clin.* 45 (2): 27–37.

21 Montoya, J., Holmes, T., Anderson, J. et al. (2017). Cytokine signature associated with disease severity in chronic fatigue syndrome patients. *Proc. Natl. Acad. Sci. USA* 114: E7150–E7158.

22 Albrecht, D., Forsberg, A., Sandstrom, A. et al. (2019). Brain glial activation in fibromyalgia - a multi-site positron emission tomography investigation. *Brain Behav. Immun.* 75: 72.

23 Gadani, S., Cronk, J., Norris, G., and Kipnis, J. (2012). IL-4 in the brain: a cytokine to remember. *J. Immunol.* 189: 4213–4219.

24 The REMAP-CAP Investigators (2021). Interleukin-6 receptor antagonists in critically Ill patients with Covid-19. *N. Engl. J. Med.* 384: 1491–1502.

25 Scaravillie, F., An, S., Groves, M., and Thom, M. (1999). The neuropathology of paraneoplastic syndromes. *Brain Pathol.* 9: 251–260.

26 Wallukat, G., Hohberger, B., Wenzel, K. et al. (2021). Functional autoantibodies against G-protein coupled receptors in patients with persistent Long-COVID-19 symptoms. *J. Transl. Autoimmun.* 4: 100100.

27 Couzin-Frankel, J., Vogel, G. (2022) In rare cases, coronavirus vaccines may cause Long-Covid-like symptoms. *ScienceInsider* (20 January).

28 Pam, B. (2022). New research hints at 4 factors that may increase chances of Long Covid. *The New York Times* (5 January).

29 Su, Y., Yuan, D., Chen, D.G. et al. (2022). Multiple early factors anticipate post-acute COVID-19 sequelae. *Cell* 185: 881–895.

30 Sneller, M.C., Liang, C.J., Marques, A.R. et al. (2022). A longitudinal study of COVID-19 sequelae and immunity: Baseline findings. Ann. Intern. Med. (Epub 24 May). https://doi.org/10.7326/M21-4905.

31 Donnelly, C., Andriessen, A., Chen, G. et al. (2020). Central nervous system targets: Glial cell mechanisms in chronic pain. *Neurotherapeutics* 17 (3): 846–860.

32 Chaves-Filho, A., Macedo, D., de Lucena, D., and Maes, M. (2019). Shared microglial mechanisms underpinning depression and chronic fatigue syndrome and their comorbidities. *Behav. Brain Res.* 372: 111975.

33 Sotzny, F., Blanco, J., Capelli, E. et al. (2018). Myalgic encephalomyelitis/ Chronic Fatigue Syndrome-evidence for an autoimmune disease. *Autoimmun. Rev.* 17: 601.

34 Loebel, M., Grabowski, P., Heidecke, H. et al. (2016). Antibodies to β adrenergic and muscarinic cholinergic receptors in patients with Chronic Fatigue Syndrome. *Brain Behav. Immun.* 201652: 32.

35 Misiak, B., Beszlej, J., Kotowicz, K. et al. (2018). Cytokine alterations and cognitive impairment in major depressive disorder: From putative mechanisms to novel treatment targets. *Prog. Neuro-Psychopharmacol. Biol. Psychiatry* 80: 177–188.

36 Galecki, P., Mossakowska-Wojcik, J., and Talarowska, M. (2018). The anti-inflammatory mechanism of antidepressants – SSRIs, SNRIs. *Prog. Neuro-Psychopharmacol. Biol. Psychiatry* 80: 291–294.

37 Sugama, S., Cho, B., Baker, H. et al. (2002). Neurons of the superior nucleus of the medial habenula and ependymal cells express IL-18 in rat CNS. *Brain Res.* 958: 1–9.

38 Pastrak, M., Abd-Elsayed, A., Ma, F. et al. (2021). Systemic review of the use of intravenous ketamine for fibromyalgia. *Ochsner J.* 21: 387–394.

39 Saligan L, Luckenbaugh M, Slonena E, et al. (2016) An Assessment of the Anti-Fatigue Effects of Ketamine from a Double blind, placebo controlled crossover study in bipolar depression. *J Affect Disord.* 194: 115–119.

Section 4

Evaluation and Management

8

Patient Evaluation and Research

Need: Widely Accepted Long-COVID Term and Case Definition

Diagnostic terms and codes should be established globally for proper documentation. The diagnostic term post-acute sequelae of SARS-CoV-2 infection (PASC) is used in research and medical articles, but the term favored by patients and the media is long COVID. These two terms should be considered interchangeable. As of the time of writing, a PubMed search yields twice as many citations for "long COVID" (800) as "PASC". The World Health Organization (WHO) has recently assigned an emergency use *International Classification of Diseases, Tenth Revision* (ICD-10) code (U09.9) referring to 'post-COVID conditions', but it is important that ICD coding be flexible with diagnostic terms. However, in clinical practice, the recognition and diagnosis of long COVID has been uneven, 27% of UK practices had never used the ICD code for long COVID as of May 2021 [1].

We have used, and for research recommend, the NIH/NICE PASC case definition for long COVID: (i) documented SARS-Cov-2 infection, (ii) duration of symptoms greater than three months, and (iii) more than three of the following symptoms: fatigue, dyspnea, musculoskeletal pain, headaches, cognitive disturbances, sleep disturbances, and mood disturbances. There should be flexibility in documenting SARS-CoV-2 infections. An analysis using these diagnostic criteria to predict the accuracy of the ICD code provided a very good predictive value greater than 90% [2]. The predictive value of these diagnostic criteria did not vary with age, sex, hospitalizations or using 12 rather than six weeks as cut-off for duration of symptoms, suggesting that the ICD-10 code for 'post-COVID conditions' is a reliable criterion for future research.

There is a general consensus that long COVID should be differentiated from the expected, persistent symptoms that follow an acute SARS-CoV-2 infection.

Unravelling Long COVID, First Edition. Don Goldenberg and Marc Dichter.

As we emphasized in Section 1, Chapter 1, long COVID should be reserved for patients that lack a medical explanation for their persistent symptoms, and as Professor David noted, "...it should not be a catch-all category where any disorder with unexplained symptoms can be attached. For example, if a patient recovers from the acute respiratory illness, but remains short of breath and is found to have pulmonary fibrosis or pericarditis by accepted criteria, or, experiences brain fog and mental slowing, later linked to microvascular infarcts on MRI – can they be removed from the post-COVID-19 cohort? I would say yes. Their condition may be unusual, and it may be serious, but it is not mysterious" [3]. This advice was not followed in many prior studies. For example, patients with post-ICU syndrome (PICS) should not be grouped with long-COVID syndrome, since their clinical manifestations and pathophysiology are so different. We used hospitalization to separate what we termed long-COVID disease from long-COVID syndrome. There are other distinguishing features that can be used, including evidence of pulmonary or cardiac damage during the initial infection.

Need: Uniform Symptom and Outcome Measures

It is essential to document the details of the acute SARS-CoV-2 infection, including dates, specific symptoms, whether the diagnosis was confirmed (testing, hospitalization) or suspected, prior health status, and the timing and severity of initial symptoms, including selected test results. A long set of symptoms were recommended to evaluate any person with acute COVID-19 [4]. The WHO has led efforts to standardize long-COVID terminology and case definitions internationally [5]. We have narrowed the initial symptom checklist to include those symptoms that are most important to measure in long COVID (Table 8.1).

The WHO has developed a post-COVID-19 condition care report, and it will be important to share such forms internationally [6]. There are outcome instruments available that measure characteristic long-COVID symptoms in other chronic medical conditions, including CFS/ME and fibromyalgia, but they will need to be modified and evaluated for long COVID. This will include standardized medical terminology. For example, loss of smell, ansomia, is different from a distorted smell, or parosmia, which need to be differentiated in studies of long-COVID symptoms.

It will also be important to agree upon common terminology when reporting long-COVID symptoms. The Human Phenotype Ontology (HPO) was structured to provide 287 HPO terms to help patient and clinician symptom integration [7]. As we have discussed, asking a patient about fatigue often fails to distinguish the mental and physical components of fatigue or its severity. Instruments used to measure fatigue include the Modified Fatigue Impact Scale, the Fatigue Severity

Table 8.1 Symptom and outcome measures in long COVID.

Symptoms
General: fatigue, weight loss, loss of appetite, malaise, weakness
Cardiopulmonary: dyspnea, palpitations, chest pain, cough, orthostatic
Endocrine: diabetes, hyperlipidemia, HPA-axis-suppression symptom
Gastrointestinal: diarrhea, constipation, abdominal pain, IBS
Neuropsychiatric: depression, anxiety, PTSD, sleep disturbance, cognitive disturbance, neuropathy, loss of taste/smell
Pain: myalgia, joint pain, headache, widespread pain (fibromyalgia)
Function
Physical: exercise, mobility, work-related
Emotional: mood, coping, fear, helpless, hopeless, stigma
Social: isolation, family, friends
Global well-being
Quality of life
Self-care, independence
Need for medical care, hospitalization, care burden
Financial impact

Scale, and the Chalder Fatigue Questionnaire [8]. Tests to differentiate central from peripheral pain include the Fibromyalgia Severity Scale and Brief Pain Inventory [8]. In addition to measuring orthostatic changes in blood pressure and pulse, standardized tests for autonomic dysfunction include the Standing Tolerance Test and the Composite Autonomic Symptom Scale 31 [9]. Common cognitive tests are the Cognitive Functioning Self-Assessment Scale and the Montreal Cognitive Assessment, instruments to measure mood include the General Anxiety Disorder Assessment-9, the Hospital Anxiety and Depression Scale, Short Form-36 (SF-36), the Posttraumatic Stress Disorder Checklist for DSM-5, the Primary Care Post-Traumatic Stress Disorder Screen, and the Yale-Brown Obsessive Compulsive Survey [9]. Standardized measures of physical function and quality of life include the Two-Minute Step Test, Six-Minute Walk Test, Activity Measure for Post-Acute Care, and the EuroQol Visual Analogue Scale [9]. Sleep instruments used in long COVID include the Insomnia Severity Index and the STOP-Bang Questionnaire [9].

We suggest neurocognitive, pulmonary, cardiovascular evaluation and treatment guidelines (Figures 8.1 and 8.2) [10]. Long-COVID patients with a high likelihood of cardiac involvement may receive blood tests (e.g. C-reactive protein (CRP), troponin, B-type natriuretic peptide/NT-proBNP), a transthoracic echocardiography, cardiac MRI, stress single positron emission CT, Holter monitor, or even coronary angiography (Figure 8.2) [11].

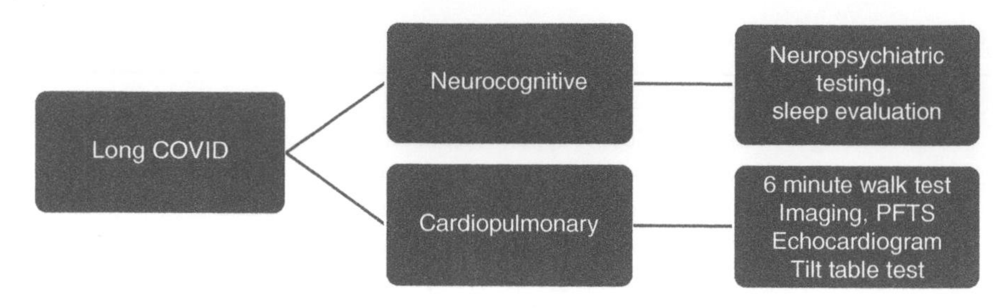

Figure 8.1 Initial long-COVID patient evaluation.

Initial Evaluation, Primary Care Role

Primary care clinicians are the frontline evaluators and care providers for patients with long COVID and their first point of medical contact. Primary care in the US is not set up well to meet the demands of long COVID. US clinical services and research are organized within disease-specific silos. The fee-for-service system rewards procedure-oriented specialty care and makes comprehensive primary care financially unsustainable. Dr. Daniel Horn, a primary care physician at Massachusetts General Hospital, said, "A high functioning primary care network like mine would operate at a net loss of 20 to 30% a year, or an average annual cost of [US]$150,000 per physician. . . That's partly because we employ diabetes educators, geriatric case managers, social workers, and addiction-recovery coaches — health professionals who are essential to the well-being of our patients but whose work is not reimbursed by insurers. Most primary care doctors in our country don't have this support system. America began this pandemic with a national primary care shortage, and without help, they now face existential peril" [12].

In the UK and other countries with universal health coverage, primary care is the center of healthcare. More than 90% of all National Healthcare Service contacts take place within primary care [13]. Primary care teams including general practitioners (GPs), advanced clinical practitioners, healthcare assistants, clinical pharmacists, and physical therapists are integrated within a team care structure in the UK. In 2020, new primary care networks (PCNs) were established in the UK and there are now 1250 PCNs serving communities of 30 000–50 000 patients [14]. These government funds provided "financial relief to primary care practices by switching from fee for service to a monthly bundled payment. . .provide all primary care practices nationwide with a reasonable fixed payment, say on average $50 per patient per month" [15].

In the UK, primary care providers have easy access to link workers, non-physician community care workers [16]. Primary care research in the UK has recognized the importance of holistic care, for example, the randomized study to

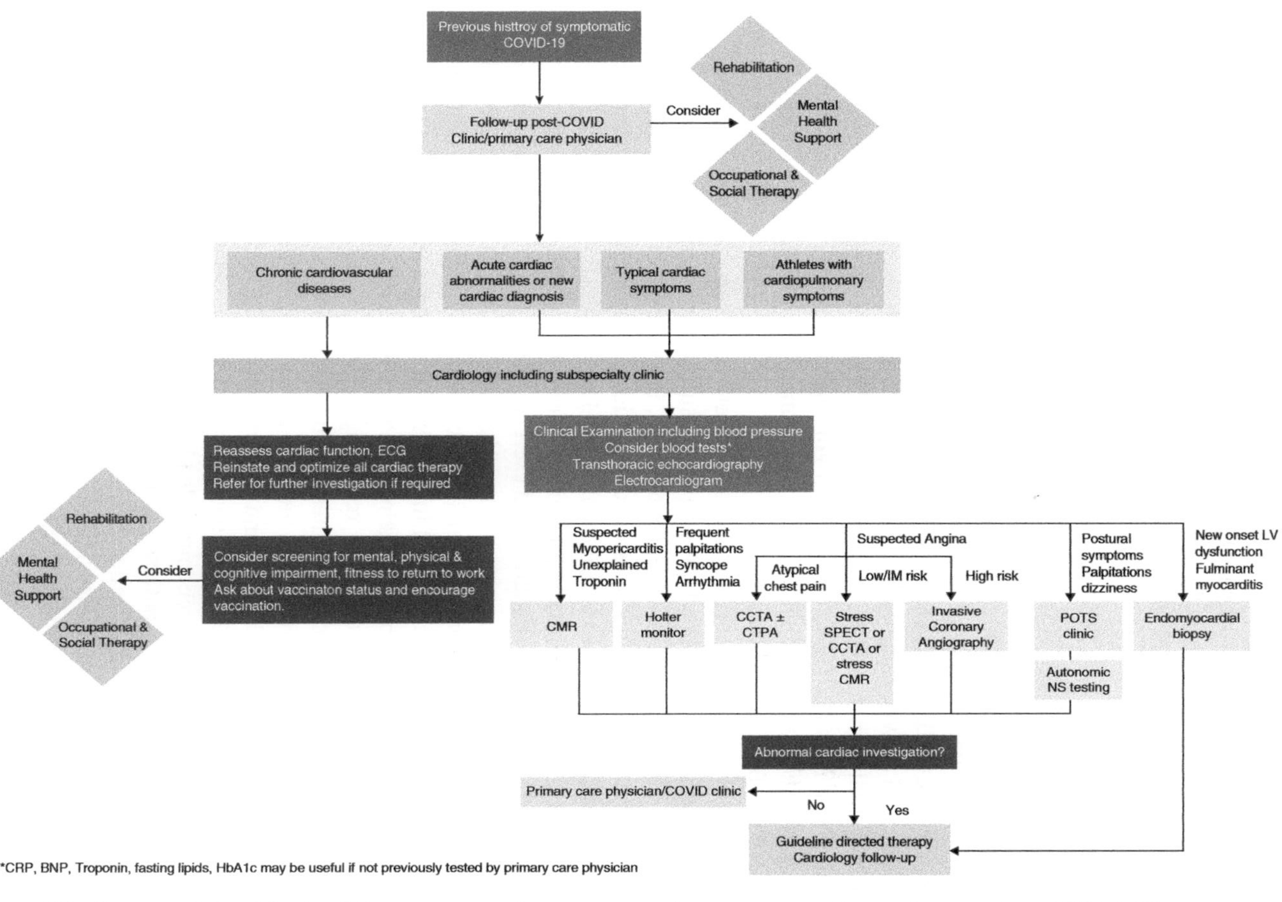

Figure 8.2 Suggested algorithm for follow-up care and management of post-acute cardiovascular sequelae of COVID-19. *Source:* With permission from Raman et al. [11]. Figure 3, p. 13.

determine the effect of the Optimal Health Programme, an evidence-based psychosocial support program [17].

The time constraints on primary care make the initial long-COVID patient evaluation extremely difficult. Before the pandemic, the average primary care clinic visit in both the US and UK was 10 to 15 minutes [18, 19]. Both countries embraced remote consultations, an appropriate, time saving means of triaging patients with long COVID. In the UK, primary care providers now offer an extended visit to potential long-COVID patients along with a set of written, detailed information [20] about the illness.

Primary care providers must be able to recognize the multiple symptoms and various presentations of long COVID and exclude medical conditions that could mimic long COVID. This evaluation should include a thorough patient history and exam as well as, when appropriate, laboratory testing. Although there are no prospective studies on initial testing, it is reasonable to obtain complete blood counts, acute phase reactants (ESR, CRP), blood glucose and kidney, liver, and thyroid function tests. In patients with cardiac symptoms, CPK-MB, D-dimer, troponin, and NT-pro-BNP, should be considered [21].

The initial evaluation must include a complete neurocognitive and cardiopulmonary assessment (Figure 8.1). A simple measure of cardiopulmonary function, such as the six-minute walk distance (6MWD), is appropriate. If dyspnea is present, the pulmonary evaluation should include chest imaging and pulmonary function studies, where the most likely abnormal test is diffusion capacity (DLCO). Chest X-rays are a poor marker for recovery from long COVID and have a low resolution [22]. An echocardiogram and exercise tolerance test should be ordered if there are concerns regarding cardiac function. ANS testing, such as the tilt table test, should be part of the initial evaluation, but a more detailed ANS evaluation may be important, as recommended in the UK, "For people with postural symptoms, for example palpitations or dizziness on standing, carry out lying and standing blood pressure and heart rate recordings (3-minute active stand test for orthostatic hypotension, or 10 minutes if you suspect postural tachycardia syndrome, or other forms of orthostatic intolerance. . .)" [23]. A global list of doctors with experience in the evaluation and treatment of POTS and ANS disorders is available at Dysautonomia International (http://dysautonomiainternational.org). Routine brain MRIs are not recommended but should be ordered in consultation with neurology [24]. Each patient should be carefully evaluated for mood and sleep disturbances. Standardized tests commonly used in general practice include the Hospital Anxiety and Depression Scale (HADS) and Short Form-36 (SF-36). If there is concern for a primary sleep disturbance, polysomnography should be ordered.

Approximately 90% of initial visits for long COVID will be in primary care [25]. The absence of any established guidelines for long COVID has fostered a fragmented, symptom-by-symptom, diagnostic approach. Specialized clinics, specialty referral systems, virtual care, and home-based care are increasingly used in the evaluation of long-COVID patients [26]. A centralized referral system for patients with long COVID would greatly expedite their care. Currently, the health care professionals most often involved in long-COVID patient evaluation are pulmonary/respiratory specialists (100%), cardiovascular specialists (92%), psychiatry and psychology (83%), physiotherapy (83%), occupational therapy (75%), social work (75%), neurology specialists (75%), primary care physicians (58%), nutrition (58%), and speech and language therapists (50%) [26].

Two UK long-COVID patients described their vastly different experiences with primary care visits. One 40-year-old female lamented, "I tend to speak to different GPs every time. There was one GP who just thought it was all anxiety. . . she said, 'There's nothing wrong with your lungs. This is all anxiety. You must treat your anxiety. There's nothing wrong with you. How are you going to manage the pandemic if you don't treat your anxiety?' That was really upsetting because I knew I was short of breath. . . I just cried and also, it didn't help because, at that stage, it wasn't known about the ongoing symptoms and my husband wasn't massively supportive at that time" [27]. A 42-year old woman described a much different GP encounter, "I have to say it was a really powerful experience speaking to the Gps. . . the two more recent ones, actually just the experience of being heard and feeling like somebody got it and was being kind about it, but you know it was okay that they couldn't do anything, I just kind of needed to know that I wasn't losing it really and it was real what I was experiencing, I think so that was really helpful" [27].

Long-COVID Clinics

It soon became apparent to long-COVID patients and their providers that specialized long-COVID care centers would be an ideal approach for both treatment and research. Patients felt that their providers did not have the time or expertise to deal with their multiple symptoms and were often referred to various specialists. One long-COVID patient said, "What we need is a service where people can get referred to, where people listen to them and then do the screening . . . to screen the people who have symptoms suspicious of cardiac and respiratory issues that need further investigation. Now, that's not to say everybody with respiratory symptoms will need further investigation . . . So a service whereby there are, where somebody who is able to use a commonsense approach, listens to people and says, 'Well, okay, you don't need any more', or 'you need further secondary care input as in cardiology or respiratory or whatever', or dermatology even, you know, 'cause there's lots of different things. And neurology" [27].

Medical providers understood that research on long COVID was desperately needed and could only be done in a federally funded program. Shortly after the US and UK allocated major funding for long-COVID research in 2021, multidisciplinary long-COVID clinics began cropping up. By February 2022, there were more than 50 long-COVID clinics in the UK and about 200 in the US with at least one clinic in almost every state (see Appendix).

One of the first long-COVID clinics was established at New York City's Mount Sinai Health System in May 2020. The clinic's director, Dr. Zijian Chen, was initially overwhelmed, "I looked at the number of patients that were in the database and it was, I think, 1800 patients. I freaked out a little bit. Oh my God, there's so many patients telling us that they still have symptoms. This was a puzzling group where we couldn't see what was wrong. These tended to be the patients who had originally had mild to moderate symptoms. They were overwhelmingly women, even though men are typically hit harder by acute COVID-19" [28]. Dr. Dayna McCarthy, a rehabilitation specialist at Mount Sinai, who works with Dr. Chen said, "Many of these patients have had million-dollar work-ups, and nothing comes back abnormal" [28].

Dr. Benjamin Abramoff, director of the University of Pennsylvania Post-COVID Assessment and Recovery Clinic, described the wide range of patient problems he was encountering, "We see patients who are going to work who have persistent loss of smell and they can't eat like they're used to, and that's going to be very bothersome and distressing. And then we have patients who are so impaired with fatigue that they can't get out of bed. What we've learned treating patients with neurologic disability is that those are total body problems. And oftentimes with patients with long COVID, there's a lot of different elements at play" [28].

Many of the long-COVID clinics associated with major medical centers have received substantial research funding from an NIH-sponsored program called RECOVER (Researching COVID to Enhance Recovery) and are tasked with finding novel approaches to prevent and treat long COVID. Awardees include the Brigham and Women's Hospital, Case Western Reserve School of Medicine, Howard University School of Medicine, Ichan School of Medicine at Mount Sinai, University of Alabama at Birmingham, University of Arizona College of Medicine, University of Illinois Hospital and Health Sciences System, University of Texas Health Science Center, University of Utah Health, West Virginia Health Services, Emory University, the University of California at San Francisco, and Stanford University.

As an example, the University of Alabama at Birmingham Long-COVID Clinic received $17 million over four years from the RECOVER-funded program. Co-principal investigator, Dr. Nathaniel Erdman, said, "What we now appreciate about long COVID, or PASC, is that this is not a singular process, but rather a collection of symptoms and syndromes. The study is designed to conduct more comprehensive testing for those individuals experiencing specific types of symptoms. This large collaborative effort will help determine the shared characteristics of those with PASC and what factors trigger distinct symptoms. With this knowledge, we will be in a much better position to determine appropriate treatment and interventions" [29]. Stanford's Post-Acute COVID-19 Syndrome Clinic will receive $15 million from the RECOVER program and plans to enroll 900 long-COVID patients over the next four years [30].

Most of the largest long-COVID clinics are multidisciplinary and led by a neurologist, pulmonologist, or physical medicine and rehabilitation specialist. In a survey of 45 long-COVID clinic in the US, more than 80% of the clinics were specifically opened for the evaluation and treatment of long COVID rather than redesigned from an existing clinic [9]. Forty percent of clinics are physical medicine and rehabilitation (PM&R), 22% pulmonology, and 16% internal medicine [9]. Neurology, PM&R, physical therapy, pulmonology, and cardiology were part of the team in two-thirds of the clinics. Initially, about one-half of clinics could see new referrals within one month and the most common eligibility for new patients was long-COVID symptoms persisting for more than one month since the initial SARS-CoV-2 infection. Some long-COVID clinics require more specific symptoms for self-referral, such as the COVID Recovery Clinic at Keck Medicine of the University of Southern California (USC), which requires either a positive test result for COVID-19 or antibody test within eight weeks of initial diagnosis as well as persistent COVID-19 symptoms, including depression and anxiety, for eight weeks or more after diagnosis without a fever. Many long-COVID clinics require a physician referral.

In the UK, more than 80 long-COVID clinics were established under the LOng COvid Multidisciplinary Consortium: Optimizing Treatments and Services Across the NHS (LOCOMOTION). However, access to these clinics was initially limited and waiting times were long. These clinics were tasked with clinical care but also had research goals, "Our research aims to produce a 'gold standard' for care by analyzing what is happening to patients now, creating new systems of care and evaluating them to establish best practice. This research is also based on the experience of a wide range of NHS professionals already treating people in ten [long-COVID (LC)] clinics across the UK and led by academics (universities) with links to other LC funded studies. The research will take place in three settings: LC clinics; at home (including self-monitoring on a mobile device using a set of questions on symptoms built into an app); and in doctors' surgeries. We will track where patients are being referred or not referred and learn from the experience of clinics by interviewing patients and recording outcomes" [10].

Three-quarters of the clinics use standardized evaluation measures and diagnostic tests although they vary greatly [9]. Common diagnostic tests included MRI, EKG, echocardiogram, pulmonary function tests (PFTs), pulse oximetry, orthostatic testing, and CRP. Only one-half of the clinics directly managed mental health needs. Clinics were interested in establishing collaborative studies as well as protocols for the evaluation of specific symptoms such as fatigue, cognitive disturbances, palpitations, and loss of taste/smell.

Ideally, primary care and long-COVID clinics work together for optimal evaluation and management. The UK COVID rapid guideline expert panel suggested: "To ensure people get the right care and support, a tiered approach could be used in which everyone gets advice for self-management, with the additional option of supported self-management if needed. People can then also be offered care from different services to match the level of their needs. This should take into account the overall impact their symptoms are having on their life and usual activities, and the overall trajectory of their symptoms, taking into account that symptoms often fluctuate and recur so they might need different levels of support at different times" [24].

Many long-COVID clinics were set up within existing PM&R programs, which are ideally suited to provide neurocognitive testing and document physical and psychological function [9]. PM&R programs have the most established expertise in long-term rehabilitation, which is critical in returning long-COVID patients to optimal recovery.

Most long-COVID clinics in the US and UK were established at large medical centers. However, there are smaller, independent long-COVID clinics, such as at the Watson Clinic's Post-COVID in Lakeland, Florida. Dr. Kathleen Haggerty, an internist, spends an hour with each new patient [31]. Several long-COVID clinics, especially the smaller ones, rely on telehealth to help expedite patient care. Dr. Brad Nieset, a family-medicine physician, runs the Benefis Health System's Post-COVID-19 Recovery Program in Montana, and initially evaluated Amber and Mike Rausch remotely. Amber said after their initial telehealth visit, "We felt comforted by learning that we know so much more about COVID and long-haul symptoms than we did at the beginning of the pandemic. I just remember [Nieset] giving us so much hope that day." [31].

There has been tremendous pressure on long-COVID clinics to demonstrate their efficacy, as noted by a UK expert, "The pressure to perform has been high, with healthcare professionals feeling that the completion and submission of outcomes and key performance indicators (KPIs) was equal to care. As such, teams have treated quantities of patients with no time to consider more subtle patterns of patient presentation" [32]. Dr. Stephen Martin, a professor at the University of Massachusetts Medical School, noted that long-COVID patients require the coordination of various specialists, many overbooked with wait lists of more than six months, "This really hits us in our Achilles' heel of health care. The American health care system really isn't set up to do this at scale" [33].

Currently, there are not adequate guidelines regarding when to refer patients to long-COVID clinics. Many clinics require patient symptoms to have persisted for more than three months. This seems reasonable, since many patients improve between one- and three-months post-SARS-CoV-2 infection. In most instances, patient referral is determined more by treatment concerns than diagnostic concerns.

Long-COVID clinics provide many patients with much needed confidence that there is a way forward. Dr. Monica Lypson, co-director of the COVID-19 Recovery Clinic at George Washington University, commented, "Meeting patients where they are at and believing them is the most important thing we do, telling them 'I hear you, I believe you, and I'm going to try to help you navigate the health care system and figure this out'" [34].

Research

In December 2020, the NIH convened a workshop to summarize the knowledge about long COVID and to "identify the causes and ultimately the means of prevention and treatment of individuals who have been sickened by COVID-19, but don't recover fully over a period of a few weeks" [35]. Dr. Francis Collins, the then-NIH Director noted, "One of the really troubling aspects of this terrible pandemic might be the lingering of this long tail of effect on people who are not able to return back to their pre-infection state, and we need to do everything we can to get answers to that. This is being taken with the greatest seriousness and moved forward at a scale that has not really been attempted before for something like this" [35]. Initial clinical studies were centered around the establishment of long-COVID clinics throughout the US in collaboration with the RECOVER program. A standardized set of examinations and tests were established with the ability to analyze 40 million electronic health records in real time.

Congress approved USD$1.15 billion over the next four years earmarked for long-COVID research. One element of this, called the NIH PASC Initiative, is the first extensive, government research dedicated to ". . .enhance our knowledge of the basic biology of how humans recover from infection and improve our understanding of other chronic post-viral syndromes and autoimmune disease, as well as other diseases with similar symptoms. Some of the initial underlying questions that this initiative hopes to answer are:

- What does the spectrum of recovery from SARS-CoV-2 infection look like across the population?
- How many people continue to have symptoms of COVID-19, or even develop new symptoms, after acute SARS-CoV-2 infection?
- What is the underlying biological cause of these prolonged symptoms?
- What makes some people vulnerable to this but not others?
- Does SARS-CoV-2 infection trigger changes in the body that increase the risk of other conditions, such as chronic heart or brain disorders?" [35].

The RECOVER program integrates long-COVID clinical care and clinical research. In addition to the 15 medical centers that received initial funding as part of the clinical consortium, New York University Langone Health was given funding for a clinical science core facility, Massachusetts General Hospital for a data resource core, the Mayo Clinic for a biorepository core, and RTI International as an administrative coordinating center. The Columbia University Irving Medical Center established a collaborative study that combined 14, long-term cohort studies involving more than 50 000 COVID-19 patients.

In the UK, more than £38 million was earmarked for long-COVID research by the UK NIHR and the UK Research and Innovation (UKRI). The UK ONS began a Coronavirus (COVID-19) Infection Survey early in the pandemic that carefully asks about persistent symptoms. The UK Post-hospitalization COVID-19 study (PHOSP-COVID), is "...a multidisciplinary consortium of clinicians and data and basic scientists aiming to create an integrated research and clinical platform to investigate the multidimensional long-term health outcomes in patients who have been admitted to hospital with COVID-19. It is set to enroll 8500 hospitalized patients and will also initiate a study of non-hospitalized subjects" (www.yourcovidrecovery.nhs.uk). Specific patient groups being studied include those with lung disease by the PHOSP-COVID Airways Diseases Working Group and non-hospitalized long-COVID patients by the University of Oxford Global COVID-19 Long-Term Follow-Up Study and others. The Oxford study has developed standardized data collection tools to document the clinical sequelae and risk factors of long-term COVID-19 consequences for up to three to five years and to facilitate international comparisons in adults and children. The PHOSP-COVID study is linked to immunology and neuropsychology studies in the UK.

Research teams at the University of Birmingham, in partnership with the Clinical Practice Research Datalink, use electronic GP records to study non-hospitalized patients with long COVID who have had symptoms for 12 weeks or longer for a major clinical digital study. Participating patients use a novel access to the digital platform to self-report symptoms as well as their quality of life and work capability. Some patients receive blood and other biological tests and wear a device that measures heart rate, oxygen saturation, step count, and sleep quality. Long-COVID patients provide input to study design and one patient said, "Each day is baby steps in terms of recovery, but almost one year on I am still battling a myriad of symptoms from memory loss to difficulties breathing, pins and needles, and immobility. I am delighted to be part of this research project, which will give hope to so many out there who are, like me, struggling with the longer-term crippling effects of this virus" [36].

Other European nations, including Italy, Germany, France, Spain, and Belgium are also involved in long-COVID research. An international prospective study of the long-term impact of severe COVID-19 infection on more than 1000 ICU survivors over the next two years has been started [37]. There are long-COVID research studies of non-hospitalized patients in the Netherlands, Belgium, and Czechia.

General Clinical Research

Long-term follow-up studies using pre-defined patient-related outcomes including quality of life, time to return to work, baseline physical activity, and cognitive and functional assessment of all long-COVID patients are needed to evaluate assessment and management interventions. These studies should always include community controls. Many of the early reports in non-hospitalized patients, particularly those from on-line support groups, consisted of mostly women, few minorities, those with higher levels of education, and mostly healthcare workers. In the future, similar studies should be representative of the general population. Clinical studies need to evaluate all potential risk factors, including age, gender, co-morbidities, and psychosocial parameters.

As discussed in the next chapter, there is some evidence that vaccinated patients are less likely to have long COVID. However, we need prospective, controlled studies on this important issue. We also need more granular research on long-COVID patients with asymptomatic initial infections. In one report, one-third of non-hospitalized, long-COVID patients never had symptoms during their initial infection.

Currently there are 7500 registered studies on COVID-19, including 1700 in the US. In total, only about 100 are specifically focused on long COVID (COVID NIH *http://ClinicalTrials.gov*).

Some general research themes include:

i) Detailed long-COVID studies: The LOng COvid Multidisciplinary Consortium: Optimizing Treatments and servIces Across the NHS (LOCOMOTION) follows 7000 long-COVID patients in GP practices.
ii) Investigating mind–body interactions: One of the most difficult research concerns about long COVID is to explain the lack of symptom correlation with usual objective disease markers. As we have discussed, this will require a better understanding of the complicated mind–body interactions. One research study, "Depression and Anxiety in Long Term Coronavirus Disease COVID-19" (DALT-COV), will look carefully at mood and external factors, such as stress and social isolation, and examine their correlation with neurotransmitter measures.
iii) Respiratory and sleep studies: Will investigate pulmonary, cardiac, metabolic, sleep, and mental health at six months, one year, three years, and five years

following SARS-CoV-2 infection. Tests will include PFTs, cardiac echocardiograms, sleep polysomnography, and several standardized mental health instruments as well as immune and genetic studies.

iv) Dysautonomia in Patients Post COVID-19 Infection (DYSCO). Study of heart rate variability at 3, 6, and 12 months, along with electrodermal activity and clinical symptoms in long COVID and ME/CFS.

v) Some specific research on cardiovascular manifestations of long-COVID include cardiac arrhythmias, long-term cardiac function outcomes, and long-term effects on the cardiovascular system: CV COVID-19 Registry (CV-COVID-19).

Basic Research

On February 19, 2022, a *New York Times* front page article entitled "How Long Covid Exhausts the Body," suggested a systems-wide approach to understanding long COVID, citing abnormal function of the pulmonary, circulatory, neurologic, and immune systems [38]. The article discussed research demonstrating that four factors increase the risk of long COVID: high levels of viral RNA early during the infection, autoantibodies, reactivation of Epstein–Barr virus, and Type 2 diabetes [39]. This research study examined these biologic factors and their clinical correlates in long COVID and compared extensive testing in patients during the acute infection and two to three months after the infection to those of healthy controls [39]. It then matched the biologic data to clinical symptoms, based on self-reported symptoms and electronic health records.

We argue that this study should serve as a research template for future studies on long COVID. These investigators evaluated the presence of multiple autoantibodies, including those seen in patients with systemic autoimmune disease (e.g. SS-A, SS-B, Jo-1, and ribonucleoprotein (RNP) antibodies) and novel antibodies against various cytokines. Some of their findings, if confirmed in longer studies, could impact the treatment of long COVID. For example, high levels of SARS-CoV-2 during the initial infection correlated with symptoms at three months, suggesting that an antiviral treatment might improve long COVID. Evidence of autoantibodies and increased inflammation at three months suggests that an anti-inflammatory treatment at that point in time would be effective. There was also some correlation between cytokines and T-cell subpopulations with certain long-COVID clinical manifestations, such as gastrointestinal (GI) symptoms. Other studies suggest that alterations of the GI microbiome correlate with long COVID [40].

The association of long COVID with ANS dysfunction has increased basic research on immune and ANS interactions. This includes research on functional autoantibodies against G-protein coupled receptors (GPCRs). One group found that functionally active autoantibodies targeting nervous-system GPCRs and renin angiotensin correlated with the clinical severity of COVID-19 [41].

There are many collaborative global research efforts to better understand the neurologic manifestations of long COVID. The COVID-19 Neuro Databank-Biobank (The NeuroCOVID Project) funded by the National Institute of Neurological Disorders and Stroke and part of the National Institutes of Health is the largest of these. It began at New York University Langone Health and includes a global specimen biobank. The Center will coordinate multiple brain imaging studies that include multiple techniques, such as fMRI, PET scans, and electroencephalograms (EEGs), and matching them with comprehensive clinical and biologic data.

Summary

It is essential that the definition of long COVID be standardized, and we believe that distinguishing long-COVID disease from long-COVID syndrome is important in such definitions. The initial patient evaluation for long COVID will continue to be centered in primary care, but specialty services and dedicated long-COVID clinics are integral to ongoing patient care. This chapter has reviewed current long-COVID patient evaluation and research, and Chapters 9 and 10 will discuss our suggestions for the future.

References

1. Walker, A.J., Mackenna, B., Inglesby, P. et al. (2021). Clinical coding of long COVID in English primary care: a federated analysis of 58 million patient records in situ using OpenSAFELY. *Br. J. Gen. Pract.* 71: e806.
2. Duerlund, L.S., Shakar, S., Nielsen, H. et al. (2022). Positive predictive value of the ICD-10 diagnosis code for long-COVID. *Clin. Epidemiol.* 14: 141.
3. David, A.S. (2021). Long covid: research must guide further management. *BMJ* 375: n3109. https://doi.org/10.1136/bmj.n3109.
4. Williamson P, Gargon L, Clarke M, et al. (2021) Core Outcome Set developers' response to COVID-19 [Internet]. Comet Initiative. https://comet-initiative.org/Studies/Details/1538 (accessed 20 May 2022).
5. World Health Organization (2021). Expanding our understanding of post COVID-19 condition: report of a WHO webinar – February 9, 2021.
6. World Health Organization (2021). WHO Coronavirus disease (COVID-19): Post COVID-19 condition.
7. Deer, R.R., Rock, M.A., Vasilevsky, N. et al. (2021). Characterizing long COVID: Deep phenotype of a complex condition. *EBioMedicine* 74: 103722.

8 Cella, M. and Chalder, T. (2010). Measuring fatigue in clinical and community settings. *J. Psychosom Res.* 15: 17–22.

9 Dundumalla, S., Barshikar, S., Niehaus, W.N. et al. (2022). A survey of PASC clinics: characteristics, barriers and spirit of collaboration. *PM&R.* 14 (3): 348–356.

10 Haroon, S. (2021). Major £2.2m research project aims to improve treatment and understanding of Long COVID. *University of Birmingham News* (18 February). https://www.birmingham.ac.uk/news/2021/major-2-2m-research-project-aims-to-improve-treatment-and-understanding-of-long-covid (accessed 20 May 2022).

11 Raman, B., Bluemke, D.A., Lüscher, T.F., and Neubauer, S. (2022). Long COVID:post-acute sequelae of COVID-19 with a cardiovascular focus. *Eur. Heart J.* 43 (11): 1157–1172.

12 Horn D. (2020). The pandemic could put your doctor out of business. *The Washington Post* (4 April).

13 (2016). Transforming primary care. *Lancet* 387 (10030): P1790.

14 Mughal, F., Khunti, K., and Mallen, C. (2021). The impact of COVID-19 on primary care. *J. Family Med. Primary Care* 10: 4345.

15 Ashwell, G., Blane, D., Lunan, C., and Matheson, J. (2020). General practice post-COVID-19; time to put equity at the heart of health systems. *Br. J. Gen. Pract.* 70: 400. https://doi.org/10.3399/bjgp20X712001.

16 Roland, M., Everington, S., and Marshall, M. (2020). Social prescribing-Transforming the relationship between physicians and their patients. *N. Engl. J. Med.* 9: https://doi.org/10.1056/NEJMp1917060.

17 Thompson, D.R., Al-Jabr, H., Windle, K. et al. (2021). Long COVID: supporting people through the quagmire. *Br. J. Gen. Pract.* 71 (713): 561. https://doi.org/10.3399/bjgp21X717917.

18 Gray, D.P., Freeman, G., Johns, C., and Roland, M. (2020). Covid 19: a fork in the road for general practice. *BMJ* http://doi.org/10.1136/bmj.m3709.

19 Tai-Seale, M., TG, M.G., and Zhang, W. (2007). Time allocation in primary care office visits. *Health Serv. Res.* 42 (5): 1871–1894.

20 Hussain, F.A. (2022). Facilitating care: a biopsychological perspective on long COVID. *Br. J. Gen. Pract.* 72: 30.

21 Townsend, L., Fogarty, H., Dyer, A. et al. (2021). Prolonged elevation of D-dimer levels in convalescent COVID-19 patients is independent of the acute phase response. *J. Thromb Haemost.* 19: 1064.

22 Wu, Q., Zhong, L., Li, H. et al. (2021). A follow-up study of lung function and chest computed tomography at 6 months after discharge in patients with coronavirus disease 2019. *Can. Respir. J.* 2021: 6692409.

23 National Institute for Health and Care Excellence (2021). COVID-19 rapid guideline: managing COVID-19. *NICE guideline NG191.*

24 Yelin, D., Moschopoulos, C.D., Margalit, I. et al. (2022). ESCMID rapid guideline for assessment and management of long COVID. *Clin. Microbiol. Infect.* https://doi.org/10.1016/j.cmi.2022.02.018.

25 Castanares-Zapatero, D., Kohn, L., Dauvrin, M., et al. (2021). Health Services Research (HSR) Brussels: Belgian Health Care Knowledge Centre (KCE). *Long COVID: Pathophysiology – epidemiology and patient needs.*

26 Decary S, Dugas M, Stefan T, et al. (2021). Care models for Long COVID: A rapid systematic review. *medRxiv.* http://dx.doi.org/10.1101/2021.11.17.21266404

27 Kingstone, T., Taylor, A.K., O'Donnell, C.A. et al. (2020). Finding the 'right' GP: a qualitative study of the experiences of people with long-COVID. *BJGP Open* 4 (5): bjgpopen20X101143.

28 O'Rourke, M. (2021). Unlocking the mysteries of Long COVID. *The Atlantic.* (8 March).

29 Pope, A. (2021). RECOVER post-COVID study to enroll participants from the Deep South. *UAB News* (25 Oct). https://www.uab.edu/news/research/item/12387-recover-post-covid-study-to-enroll-participants-from-the-deep-south (accessed 20 May 2022).

30 Goldman, B. (2021). Stanford medicine to enroll 900 in NIH-funded long-COVID study. *Stanford Medicine News* (22 Nov). https://med.stanford.edu/news/all-news/2021/11/long-covid-research-initiative.html (accessed 20 May 2022).

31 Caceres, V. (2021). How do post-COVID care clinics help Long COVID patients? *US News & World Report* (26 October).

32 Hussain, F.A. (2022). Life and times facilitating care. *Br. J. Gen. Pract.* https://doi.org/10.3399/bjgp22X718181.

33 Morris, A. (2021). Another struggle for long Covid patients. *The New York Times* (27 October).

34 Cooney, E. (2020). Explanations for 'long Covid' remain elusive. For now, believing patients and treating symptoms is the best doctors can do. *STAT* (29 December).

35 Collins, F.S. (2021). NIH launches new initiative to study "Long COVID." NIH (23 February). https://www.nih.gov/about-nih/who-we-are/nih-director/statements/nih-launches-new-initiative-study-long-covid (accessed 20 May 2022).

36 University of Birmingham News (2021). Major £2.2m research project aims to improve treatment and understanding of Long COVID. https://www.birmingham.ac.uk/news/2021/major-2.2m-research-project-aims-to-improve-treatment-and-understanding-of-long-covid-1 (accessed 20 May 2022).

37 Wildi, K., Li Bassi, G., Barnett, A. et al. (2021). Design and rationale of a prospective international follow-up study on intensive care survivors of COVID-19: The long-term impact in intensive care survivors of coronavirus disease-19–AFTERCOR. *Front. Med.* 8: 738086.

38 Keller, J. (2022). How long Covid exhausts the body. *The New York Times* (19 February). https://www.nytimes.com/interactive/2022/02/19/science/long-covid-causes.html (accessed 20 May 2022).

39 Su, Y., Yuan, D., Chan, D.G. et al. (2022). Multiple early factors anticipate post-acute COVID-19 sequelae. *Cell* 185 (5): P881–P895.e20.

40 Vestad, B., Ueland, T., Lerum, T.V. et al. (2022). Respiratory dysfunction three months after severe COVID-19 is associated with gut microbiota alterations. *J. Intern. Med.* https://doi.org/10.1111/joim.13458.

41 Dotan, A., David, P., Arnheim, D., and Shoenfeld, Y. (2022). The autonomic aspects of the post-COVID 19 syndrome. *Autoimmun. Rev.* 21 (5): 103071.

9

Patient Management and Rethinking Healthcare Amid Long COVID

Management

General Issues

As of February 2022, only 40% of long-COVID patients had received any specific therapy for their symptoms [1]. Three-quarters of those were for a prescribed medication. Two-thirds of hospitalized long-COVID patients, compared to one-third of non-hospitalized patients, had received any treatment. Specialists involved in treatment decisions have included pulmonologists (63%); cardiologists (55%); neurologists (30%); ear, nose, and throat (ENT) specialists (22%); physical medicine and rehabilitation (PM&R) specialists (16%); gastroenterologists (15%); psychiatrists (9%); and infectious disease specialists (8%) [1].

Long-COVID patients felt that the primary cause of their unmet care needs was a lack of information, identified by more than 50% of patients (Figure 9.1) [1]. A lack of knowledgeable staff and long wait times were mentioned by more than 20% of the long-COVID patients. Most long-COVID patients want to stay informed about new scientific findings, but they are not readily available, aside from patient support groups. One idea favored by patients was a website maintained with medical expert guidance and review.

In surveys of long-COVID patients who were not being seen in specialized long-COVID clinics, more than half of the patients described the information they received about long COVID as insufficient, including 19% who did not receive any information (Figure 9.2) [1].

Unravelling Long COVID, First Edition. Don Goldenberg and Marc Dichter.

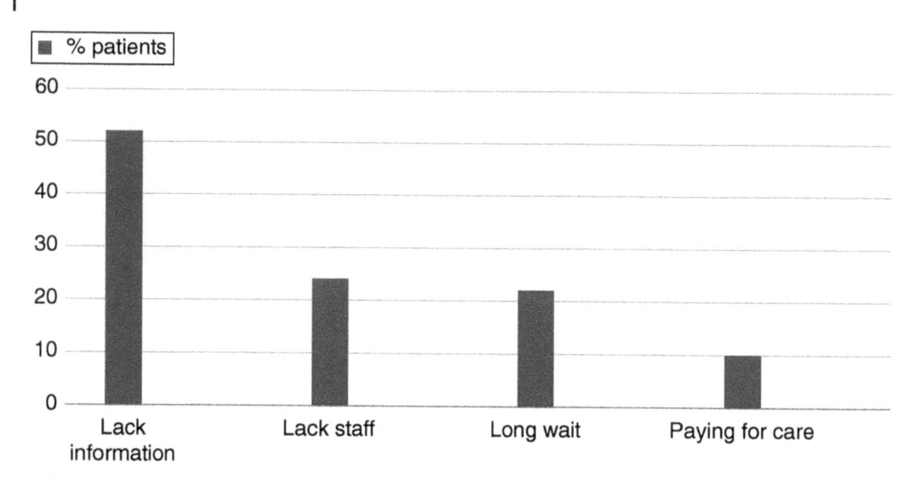

Figure 9.1 Causes of unmet care need identified by long-COVID patients. *Source:* Modified from Figure 10 [1].

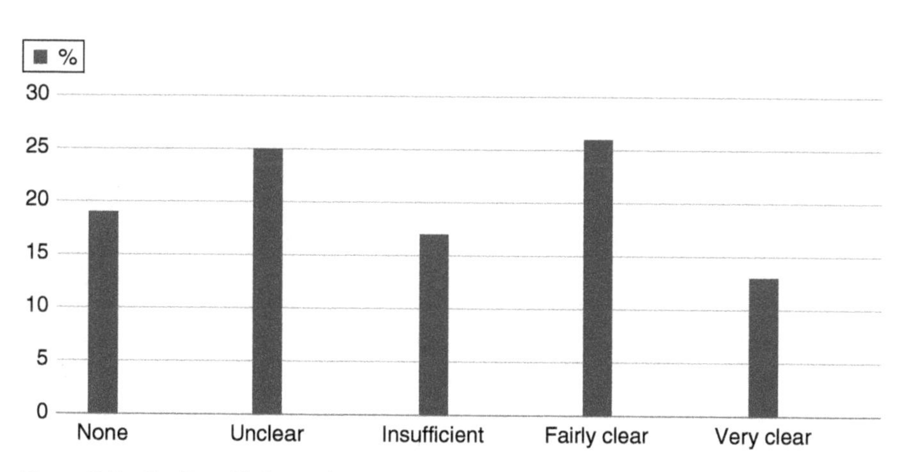

Figure 9.2 Quality of information received by long-COVID patients. *Source:* Modified from Figure 12 [1].

Pulmonary and Physical Rehabilitation

Any long-COVID patient with persistent dyspnea should have a comprehensive pulmonary evaluation. Pulmonary rehabilitation is an important aspect of most patients' management and is defined as "A multidisciplinary intervention based on personalized evaluation and treatment which includes, but is not limited to, exercise training, education, and behavioural modification designed to improve the physical and psychological condition of people with respiratory disease" [2].

Typically, this occurs under the direction of a pulmonologist and includes a skilled respiratory therapist. Many pulmonary therapists and physical therapists follow the rule-of-tens approach to guide a gradual increase level of activity. This consists of a 10% increase in duration, frequency, and intensity of the activity or exercise every ten days, as tolerated.

Pulmonologists have noted that long-COVID patients' breathing is off, ". . . long-COVID patients were breathing shallowly through their mouths and into their upper chest. By contrast, a proper breath happens in the nose and goes deep into the diaphragm; it stimulates the vagus nerve along the way, helping regulate heart rate and the nervous system. Many of us breathe through our mouths, slightly compromising our respiration, but in patients with post-acute COVID syndrome, lung inflammation or another trigger appeared to have profoundly affected the process" [3]. Dr. David Putrino and his team at the long-COVID clinic at Mt. Sinai Hospital in New York, use a twice-daily breathwork rehabilitation program called Stasis that trains people to inhale and exhale through the nose for prescribed counts [3].

A study of 58 patients who completed a 6-week pulmonary rehabilitation program included 22 hospitalized and 36 non-hospitalized patients [4]. The rehabilitation consisted of three- to four-hour, individualized endurance, strength, and inspiratory muscle training sessions three times weekly as well as patient education, psychosocial counseling by a psychologist, nutritional education, and smoking cessation sessions. These patients began their pulmonary rehabilitation an average of 4.4 months after their initial SARS-CoV-2 infection. The average age was 47 years, and 43% were female. The primary study end point was a significant improvement in the 6MWD. On average, the patients improved their 6MWD by twofold and also improved their maximal exertion (Borg scale), quality of life and fatigue scores, and pulmonary function tests (PFTs), including diffusing capacity of lung for carbon monoxide (DLCO) and FEV1.

PM&R specialists and physical therapists should be involved in the care of patients with long COVID. They will provide guidance to an individualized, symptom-guided program. Depending on the long-COVID symptoms, specialists in olfactory recovery, smell therapy, speech, and vestibular therapy may be helpful. The most difficult and controversial issue is related to exercise and cardiopulmonary rehabilitation.

Exercise

The controversy regarding the role of exercise in CFS/ME, as discussed in Chapter 5, is playing out again in long-COVID discussions. Dr. Trish Greenhalgh, professor of primary care at Oxford University and one of the clinicians who first drew awareness to the importance of long COVID, commented on exercise and the PACE trial at a public webinar on long COVID in Canada; CFS/ME advocates rebuked her saying, "Dr. Greenhalgh is a patient safety threat to all Canadians living with ME and long COVID" [5].

There is no question that patients with long COVID have extreme difficulty with physical exertion. One patient said, "Well, one of the things that really bugged me about it was the talking about graded exercise and I've learnt from experience that pushing myself even a tiny bit has massive consequences . . . I did more than I should've done, which is still probably only 25% of what I would've done before I was unwell. And was absolutely floored by it. By Monday I was in bed, I could not get out of bed and with chest pains. So to talk about graded exercise and not acknowledge the post-exertional malaise . . ." [6].

We are all aware of the physical and psychological benefits of exercise, including on our immune system. Exercise has a positive impact on depression, chronic pain, and sleep disturbances and inhibits the expression of inflammatory cytokines such as IL-6, IL-1β, and tumor necrosis factor (TNF)-α. In patients with fibromyalgia, a gradual-incremental cycling program improved centralized pain and reversed some of the functional connectivity abnormalities noted on brain imaging [7]. Increased cardiovascular fitness increases brain volume and communication between brain regions that regulate pain, cognition, mood, and sleep [8]. Moderate exercise was found to improve fatigue in fibromyalgia patients [9].

A few studies demonstrated that gradual physical activity improved pulmonary function and symptoms in patients hospitalized with severe COVID-19 [10]. In a study of nearly 50 000 patients with COVID-19, only 10% hospitalized, the mortality rate was two and a half times greater in patients who were inactive compared to patients who exercised regularly [11]. The mortality risk for being consistently inactive exceeded the odds of smoking and all the chronic diseases studied.

However, exercise must begin carefully in patients with long COVID. Dr. David Systrom, a pulmonary and critical care physician at Brigham and Women's Hospital in Boston, cautioned that long-COVID patients, "are very, very different from normal and simple detraining. There are both patients and doctors who are vehemently against any exercise. If you can get the patient in a better place with medications, then you can embark on a graded exercise program without precipitating crashes" [12] Dr. Systrom and other investigators suggest that the exercise intolerance in patients with long COVID, as in CFS/ME and fibromyalgia, is often linked to ANS dysregulation.

Dr. Sally Singh, a professor of pulmonary and cardiac rehabilitation at the University of Leicester, noted that before beginning an exercise program with her long-COVID patients, she was "aware of the rhetoric around exercise making people worse, and it was an absolute priority that we didn't do that, which is why we added in a measure of fatigue when monitoring outcomes" [5]. In 30 long-COVID patients who participated in her individually tailored, six-week rehabilitation program all except one patient improved in both energy and exercise capacity.

Dr. Paul Garner discussed the dilemma he faced when thinking about the pivotal role of exercise, "I think it is really important not to emphasize post-exertional malaise as if it is a disease. Early on I got suckered into it as something that might never go away. Part of my recovery has been around changing my thoughts around different body signals. If you see any signal as abnormal you feel insecure or get stressed, the most minor feelings get exaggerated by your brain, and you take to your bed. They believe the disease lasts for life. They reject any research that examines psychological approaches to treatment or that evaluates the role of progressive physical activity in recovery in ME/CFS, and I would assume by extension to long COVID" [5]. Dr. Garner also commented that he had "looked down the barrel of the ME/CFS gun and disarmed it," an analogy and stance that the CFS/ME community soundly rejected [5].

Tai chi and yoga have been used effectively in patients with CFS/ME, fibromyalgia, and chronic low back pain [13]. Tai chi and yoga each employ gentle, flowing body movement with deep breathing and relaxation. There are new studies investigating the efficacy of yoga, tai chi, and meditation as treatment modalities for patients with long COVID.

Evaluate and Treat Autonomic Dysfunctions and Postural Orthostatic Tachycardia Syndrome

Long-COVID patients with ANS dysfunction, especially POTS, have significant difficulty returning to normal activities. If there is evidence of ANS dysfunction, standard physical rehabilitation is likely to exacerbate long-COVID symptoms and will not be tolerated [14]. Therefore, a very gentle and individualized exercise program should be initiated. Short bouts of cardiovascular exercise while lying down or seated with the patient wearing compression stockings will reduce blood pooling. Isometric exercise should be geared toward increasing the venous return to the heart and increasing blood pressure. These might consist of tensing the thigh and buttock muscles, crossing arms and legs, folding arms and leaning forward, squatting, or raising a leg on a stool. Patients should hydrate well, at least 3 l/day, and take salt supplements. Salt intake should approach 10 g/day in adults. Prolonged standing, heavy meals, and alcohol may increase POTS. Patients are advised to sleep in a head tilt-up position of greater than 10°. Certain medications such as norepinephrine reuptake inhibitors and tricyclics may increase orthostatic intolerance and these have been used for long COVID. If hypovolemia is present, fludrocortisone can be helpful. Other medications often used include midodrine, clonidine, methyldopa, and propranolol but are best tried under the supervision of an ANS specialist.

Psychological Rehabilitation

The cognitive and psychological symptoms of long COVID are difficult to treat. Dr. Tamara Fong, a cognitive neurologist, thought that the brain fog and depression of long COVID were similar to her experience treating patients with post-concussive syndrome [15]. She treats long COVID with good sleep hygiene, stress reduction, yoga, tai chi, and meditation and noted, "Mindfulness helps. Doing too much too fast is like trying to run a marathon without training" [15]. Dr. Alexandra Merlino, a speech-language expert at the Post-COVID Assessment and Recovery Clinic at the University of Pennsylvania asks her patients with cognitive difficulty to listen to a podcast and summarize it or to memorize a grocery list, "We teach them tools like association and categorization to remember items. Many of these people have never had memory or organization problems before but suddenly they need to function in the here and now" [15].

The fear, uncertainty, and stigmatization felt by long-COVID patients has promoted anxiety, depression, sleep disturbances, and post-traumatic stress disorder (PTSD). A significant increase in mood and sleep disorders was noted in every previous coronavirus epidemic. For example, one year after Middle East Respiratory Syndrome (MERS), 27% of patients had depression and 42% had PTSD [16].

Patient and family education, including making or confirming the diagnosis, explaining the nature of the disorder, and the rationale for the treatment, are essential for optimal management of any chronic illness, especially one poorly understood. Information packets on long COVID should be made widely available.

Cognitive behavioral therapy (CBT), especially when integrated with patient education, is helpful in treating fibromyalgia and other chronic pain conditions as well as CFS/ME [17]. There is also evidence that CBT normalizes pain-related brain hyperactivity during imaging studies [18]. There are cognitive and psychoeducation clinical trials of long COVID both in the United States and United Kingdom.

It is likely that COVID-19 patients with persistent mood disturbances will benefit from CBT designed to increase self-awareness and improve coping strategies in dealing with chronic illness. The methods often used with CBT include:

- relaxation techniques,
- goal setting,
- problem solving,

- self-reinforcement, and
- substituting maladaptive thoughts with positive cognitions.

Any long-COVID patient with sleep disturbances will benefit from sleep hygiene instruction. Common things to avoid include napping during the daytime, clock-watching in bed, a noisy environment, reading or watching stressful things before bedtime, eating, or drinking before bedtime, any caffeine after noon, and avoiding alcohol. Maintaining a regular bedtime and awake time schedule is important. Medications should not be a first-line therapy and, if used, should always be combined with non-pharmacologic management. Melatonin receptor agonists, melatonin, antihistamines, and low-dose antidepressants are reasonable initial choices.

Acupuncture is used to treat fibromyalgia, chronic low back and neck pain, complex regional pain syndrome, migraine headaches, and osteoarthritis. In most studies, acupuncture results in moderate pain reduction, is better than sham acupuncture, and can improve altered brain connectivity.

Medications and Other Therapies

Currently, there are not any completed, controlled studies for the treatment of long COVID. Extrapolating from studies of similar conditions, such as CFS/ME and fibromyalgia, and the potential pathophysiologic pathways of long COVID, suggests some medications that can be used in selected patients [19, 20]. These medications may decrease pain and improve mood or sleep disturbances and include duloxetine, pregabalin, gabapentin, amitriptyline, and certain other antidepressants. Sometimes they are combined for maximum benefit. Medications that have been used to treat fatigue in neurologic disorders, such as MS, include amantadine, modafinil, and methylphenidate and are being evaluated in long COVID.

In many patients with persistent symptoms, we recommend that medications be prescribed in conjunction with randomized, clinical trials, ideally within a long-COVID clinic. This is especially important with immunosuppressive medications because of their complexity and potential toxicity. These might include plasmapheresis, intravenous immunoglobulin (IVIg), and biologic drugs. In the UK RECOVERY trial, more than 40 000 patients at 180 hospitals are being enrolled to study medications in both acute and long COVID such as monoclonal antibodies, corticosteroids, other anti-inflammatory medication, colchicine, and various immunosuppressive and biologic agents, including baricitinib and TNF inhibitors.

Other medications being studied as treatments for long COVID:

- Leronlimab (PRO 140), a humanized monoclonal antibody to the C–C chemokine receptor type 5 (CCR5).
- Montelukast, a leukotriene antagonist.

- RSLV-132, an enzymatically active ribonuclease.
- Sodium Pyruvate nasal spray.
- Cannabidiol and naltrexone, for pain reduction.
- Ampion, a low molecular weight filtrate of human serum albumin.
- Fluvoxamine and vortioxetine, two antidepressants that may affect mood, cognition, and smell.
- Statins.

There is no strong evidence that specific diets, herbal products, supplements, or vitamins are helpful in long COVID, although many people endorse them. It is possible that fortified foods, supplements, and specific diets or compounds may have a nutritional and pharmaceutical effect. Some of the nutrition-related items discussed for treating long COVID include diets high in whole grains, polyphenol-rich vegetables, omega-3 fatty acid-rich foods, and a low histamine diet with limited cheeses, fruit, seafood, and nuts. Ongoing clinical trials include:

- probiotics,
- niacinamide and omega-3,
- coenzyme Q10, and
- the roots of *Rhodiola rosea* and *Eleutherococcus senticosus*.

Additional ongoing clinical trials include treating persistent smell and taste symptoms with oral and intranasal corticosteroids, theophylline, sodium citrate, and a d *N*-methyl D-aspartate antagonist (caroverine), using acupuncture and/or transcranial, direct-current stimulation to treat pain and mood disturbances, as in depression and fibromyalgia [21], and the use of hyperbaric oxygen and vagus nerve stimulation.

SARS-CoV-2 Vaccine

In several uncontrolled reports, about one-quarter of patients with long COVID felt better two weeks to one month after vaccination. Laura Gross experienced long-COVID symptoms for months, which she described as a "zizzy-dizzy-weaky thing that was like an internal head achy all-over-body vibration, hopeless, sad, lonely, unmotivated, and brain fog barely describes it. It's more like brain cyclone" [22]. After her vaccine, "It was like a revelation, like the old me. It's like my cells went kerflooey last year when they met COVID and then the vaccine said, 'Wait, you dopes, that isn't how you fight this, do it this way'" [22].

In a self-report survey of 900 long-COVID patients, 58% reported improvement in their symptoms after COVID-19 vaccinations whereas 18% reported a worsening of their symptoms, and the rest reported no change [23]. In the largest report, including more than 2.5 million vaccinated subjects, vaccination slightly reduced the risk of long COVID by about 15% [24].

Longer and better studies are necessary to know if vaccines are indeed helpful, other than in preventing recurrent COVID-19 infection. Dr. Adam Lauring, a virologist and infectious disease physician at the University of Michigan, considered how vaccines might impact long COVID, "They might be different disease processes and you manage them differently. It might be that there's a subset of people who have a certain type of long COVID, who respond well to vaccines, but there might be other people who have a different subtype that we haven't quite defined yet" [22]. Dr. Eric Topol, an immunologist, cautioned, "We'd really like objective metrics that show that you not just feel better. You could feel better from the placebo effect, but it's unlikely your heart rate's going to go from 100 to 60 because of a placebo effect. And if we keep seeing that pattern, that would be like Eureka. I think there's probably something there, but I just don't know what is the magnitude, how many people are going to benefit" [22].

Health Coverage, Disability

Since 2021, more than 16000 Americans have filed for disability benefits related to COVID-19. Like patients with other invisible illnesses, such as CFS/ME and fibromyalgia, long-COVID patients often find it difficult to qualify for benefits. An NBC Nightly News segment on February 27, 2022, said that more than two-thirds of the long-COVID patients they surveyed were unable to obtain disability coverage for their condition [25]. One study in the United Kingdom found that one-quarter of employers cited long COVID as the leading cause of work absenteeism [26]. Health economists have estimated that one-third of the COVID-19 health burden will be related to COVID-19-induced disability [27].

Patients with long COVID face an uphill fight for obtaining any disability rights. First, as is the case for many hidden disabilities, it is difficult to demonstrate that long COVID causes physical problems that interfere with work ability. Furthermore, the Social Security Administration requires people to have been disabled for at least one year before receiving compensation. In July 2021, President Biden promised "to make sure Americans with long COVID who have a disability have access to the rights and resources that are due under the disability law" [28].

An updated government long-COVID disability guidance stated, "A person with long COVID has a disability if the person's condition or any of its symptoms is a 'physical or mental' impairment that 'substantially limits' one or more major life activities. This guidance addresses the 'actual disability' part of the disability definition. The definition also covers individuals with a 'record of' a substantially limiting impairment or those 'regarded as' having a physical impairment (whether substantially limiting or not). This document does not address the 'record of' or 'regarded as' parts of the disability definition, which may also be relevant to claims regarding long COVID. An individualized assessment is necessary to determine whether a person's long COVID condition or any of its symptoms substantially limits a major life activity" [29].

Rethinking Healthcare Amid Long COVID

Dealing with Medical Uncertainty

Long-COVID syndrome is full of uncertainty. Phillips and Williams said, "Long COVID is not a condition for which there are currently accepted objective diagnostic tests or biomarkers. It is not blood clots, myocarditis, multisystem inflammatory disease, pneumonia, or any number of well-characterized conditions caused by COVID-19. No one knows what the time course of long COVID will be or what proportion of patients will recover or have long-term symptoms. It is a frustratingly perplexing condition. There is currently no clearly delineated consensus definition for the condition; indeed, it is easier to describe what it is not than what it is" [30].

However, there is nothing unique about such medical uncertainty. Conditions with chronic, unexplained symptoms like CFS/ME, fibromyalgia, and POTS, have been referred to as medicine's blind spot, "long COVID might exemplify the category of mysterious, unexplained, chronic symptoms (either post-infectious or not), and could operate via similar mechanisms to symptoms seen in other patients. The major problem in teasing this out is that the latter, despite a long history of high numbers of patients, has remained very poorly understood and constitutes one of medicine's largest blindspots." [31].

Long COVID, like CFS/ME and fibromyalgia, lacks defined disease markers or pathologic findings. Diagnosis and treatment are based on trial and error. Dr. David Putrino, Director of Rehabilitation at Mount Sinai Medical Center said, "A lot of clinicians want the algorithm. There is no algorithm. There is listening to your patient, identifying symptoms, finding a way to measure the severity of the symptoms, applying interventions to them, and then seeing if those symptoms resolve. That is the way that medicine should be" [3]. Dr. Dayna McCarthy, who works with Dr. Putrino, said, "Everything was coming back negative, so of course Western medicine wants to say, 'You're fine'" [3].

When physicians do not have the answers, we often resort to innuendo or half-truths. Rather than admit that we do not know the structural cause of chronic low back pain in 90% of people, we diagnose slipped disks or a spinal or sacroiliac misalignment. Dr. Lewis Thomas summed up human uncertainty, "The only solid piece of scientific truth about which I feel totally confident is that we are profoundly ignorant about nature. We do not know how humans 'work' or how they 'fit in' to the enormous, imponderable system of life in which we are embedded as working parts" [32].

A rheumatology colleague in New York, Dr. Michael Lockshin, spent his career investigating the immune mysteries of systemic lupus erythematosus (SLE) and related conditions but was cognizant of their uncertainty, "I now accept that uncertainty occupies a substantial part of the world in which my patients, my students, and I live. We do not need to hide it from view. It is not cause for fright. Uncertainty is just another tool that we can learn to use" [33].

Patient Advocacy

Long-COVID patients were not getting answers from their healthcare professionals, and often felt stigmatized. One patient lamented, "I had been dismissed and turned down and completely gaslighted by doctors for months" [3]. Another patient reported, "Every specialist I saw—cardiologist, rheumatologist, dermatologist, neurologist—was wedded to this idea that 'mild' COVID-19 infections last two weeks. It was incredibly difficult because we were not believed. It's one thing to have the physical pain, but it's another to have those pains dismissed because someone has made up their mind that that's not real, that that's not attributed to COVID. That hurt worse." [34].

There is a disconnect between long-COVID patients' lived experiences and the perceptions and advice from healthcare professionals. Patients report difficulty accessing long-COVID care. "This led to patients describing how they felt they had to manipulate the inflexible algorithm-driven systems in order to receive care, which led to feelings of guilt and anger. Some patients described creative solutions they had come up with to help them access healthcare, while others reported resorting to private healthcare to access tests. Many patients felt they needed to conduct their own research and construct their own care pathways, taking the lead in arranging consultations with specialists and circumventing bottlenecks in the system" [35].

The lack of medical guidance for long COVID helped fuel intense global patient activism and resulted in patients becoming increasingly involved in their care. Patient support groups exist for every chronic illness identified during the past 20 to 30 years, but nothing close to the magnitude or impact as that of long COVID. In the past, patients were typically not involved in healthcare measures, but that has been slowly changing. Long COVID was called "the first illness created through patients finding one another on Twitter" [36]. Traditionally, diseases are described and named by clinicians and scientists according to defined criteria. The long-COVID term was first used by Elisa Perego, a COVID-19 patient from Lombardy, Italy, after struggling with her multiple persistent symptoms [37].

Long-COVID groups on social media spread first in Europe, then throughout the United States. Many of these were spearheaded by healthcare professionals suffering from long COVID. In September 2020, a group of 39 physicians suffering from long COVID published "From doctors as patients: a manifesto for tackling persisting symptoms of COVID-19" [38]. The article began, "We write as a group of doctors affected by persisting symptoms. We aim to share our insights from both a personal experience of the illness and our perspective as physicians . . . lessons learnt from other illnesses have shown the importance of involving those most affected. Patients experiencing persisting symptoms of covid-19 have a great deal to contribute to the search for solutions. Involving patients in research design and the commissioning of clinical services will ensure that the patient perspective is listened to and will optimise the development of such studies and clinical services" [38]. This article was especially insightful as healthcare professionals became long-COVID patients and discussed facing the same doubts and skepticism of other patients, which validated the long-COVID patient experience.

What is unique about long-COVID patient groups is the incorporation of sophisticated data-gathering research alongside patient support and advocacy. Support groups such as Body Politic, Long COVID Support Group, and Survivor Corps were instrumental in broadening long-COVID clinical research. The Body Politic Covid-19 Support Group has more than 10000 members from more than 30 countries (wearbodypolitic.com). One early study designed and reported by COVID-19 patients enrolled 3762 patients from 56 countries [39]. Such web-based studies completed by patients have provided greater detail about pre-existing and new post-COVID-19 symptoms.

Long-COVID online communities in Spain were established jointly with the Spanish Society of General Practitioners [40]. In France, multiple social media groups coordinated their advocacy actions [41].

Tension has always existed between patient support groups and clinicians. There is concern that support groups' may not be scientifically objective. The most zealous support groups are characterized as demanding, expressing hostility toward healthcare professionals, and feel that they have to fight for every available treatment. Support groups, by their very nature, self-select for patients who are likely not doing well and may not be representative of an illness in the general population. Patients sometimes feel overwhelmed in support groups and find it difficult to dissect truth from fantasy. Comments from long-COVID patients include, "Initially it was very encouraging that I wasn't on my own, and I wasn't making it all up and I did not feel like I was the only one, but the more it's gone on, it's scary, so I try to avoid it now, it affects my mental health, and I do not feel like I resonate any more with some of the stuff", and "Internet support groups, yeah on the Facebook groups that I'm on, I mean to be honest, I try not to read that group too much because it depresses me, makes me a bit anxious" [6].

Sharing medical information on social media often results in patients trying untested therapies, "I mean initially I started taking vitamin D. Had a joint vitamin C and zinc thing, which I didn't take every day but I took some multivitamins, but then I was a bit unsure really. . . I hope the acupuncture and the massages, the deep tissue massage and the dry-needling will help, but yeah I mean you never know, what I found helpful for me with the fatigue was anti-inflammatory but also anti-histamine diet—it's very strict, but it really helped me, so the moment I started to do it, I noticed improvements with the neurological symptoms as well" [6]. Dr. Garner, in his bout of long COVID, reported that he needed to quit "the constant monitoring of symptoms and avoided reading stories about illness and discussing symptoms, research, or treatments by dropping off the [long-COVID] Facebook groups" [42].

Social scientists have described the ways that long COVID has provided a framework for incorporating subjective online patient experience into illness recognition and identity, beyond the conventional channels of medicine [43]. A long-COVID patient voiced concern about incorporating patients' lived illness experience with observed medical data, "We're still prioritizing the medical voice over the patient voice. And you have to have both. That's where medical innovation can occur. But right now, it's not happening that way and that's been a bit of a challenge for us" [44].

The experiences during long COVID demonstrate the value of co-designing medical care and healthcare research with the patients [45]. This requires a partnership and shared leadership with healthcare providers and patients applying their unique perspectives. This does not mean that patients should become research design experts, but that they should take the lead in providing their lived clinical experience. Patients and their healthcare professionals must share decision-making. Since long-COVID therapy is largely trial and error, provider and patient dialogue and flexibility must be ongoing. Didactic dogma is not productive. In the United Kingdom, the NICE guidelines have integrated patient groups, professional societies, and medical centers. The pandemic and long COVID made it clear that healthcare providers must consider our patients' lived experience with illness. Identifying the correct diagnosis and understanding disease pathophysiology are our jobs but we cannot put ourselves in our patients' shoes. No one can speak for our patients other than themselves.

Long-COVID organizations include:

- COVID Advocacy Exchange (www.covidadvocacyexchange.com),
- National Patient Advocate Foundation COVID Care Resource Center (https://www.patientadvocate.org/covidcare),

- Body Politic COVID-19 Support Group (https://www.wearebodypolitic.com/covid19),
- Survivor Corps (https://www.survivorcorps.com), and
- Patient-Led Research for COVID-19 (http://patientresearchcovid19.com).

Integrated, Patient-centric Care

Remember our patient Sarah who first saw her primary care physician for symptoms of long COVID and over the next few months consulted with a pulmonologist, cardiologist, neurologist, endocrinologist, immunologist, and a psychiatrist. Patients with long COVID were seeing many physicians, each focusing on their organ system specialty. This is common in the US health system, which is divided into specialty silos.

Integrated, patient-centric care is the optimal way to manage any chronic illness, especially a complicated disorder such as long COVID. Patient-centeredness is defined as "Health care that establishes a partnership among practitioners, patients, and their families. . . to ensure that decisions respect patients' wants, needs, and preferences and that patients have the education and support they need to make decisions and participate in their own care" [46]. Patient-centric care is not focused on technology, disease, doctors, or hospitals, but on the patient. This does not mean that every patient wants to share in every bit of information and decision-making. That balance should be determined individually.

Such integrated care revolves around the patient and the primary healthcare provider, which is not easy to sustain in the United States where healthcare is a patchwork of primary and specialty services with little coordination. Primary care in the United States, including family practice, internal medicine, and pediatrics, is undervalued and underpaid. Dr. Donald Berwick said, "Despite spending on health care being nearly double of the next most costly nation, the United States ranks 31 on life expectancy, 36 on infant mortality . . . As a side effect of the cost burden, the United States is the only industrialized nation that does not guarantee universal health insurance to its citizens. We claim we cannot afford it [47].

Our US fee-for-service reimbursement system rewards healthcare providers for more services, particularly technical procedures, without accounting for value. When countries increase their investment in primary care, it lowers healthcare costs, improves health, and decreases mortality [48].

Optimal chronic care management should also be based on incremental care, that is, a care system that evolves gradually over time, rather than episodically. Dr. Atul Gawande summed up our current care system, "We pay doctors for quantity, not quality . . . we also pay them as individuals, rather than as members of a team working together for their patients. Both practices have made for serious problems" [49].

Integrated, stepped care must recognize the importance of mental health evaluation and treatment. Primary care and community health centers should include mental health workers as part of the care team. Currently, 80% of antidepressant and antianxiety medications in the United States are prescribed by primary care doctors, despite their lack of mental health training. The dramatic shift in mental health care from specialty care to primary practice reflects the lack of funding for mental health nationally. In integrated, primary, collaborative care, a nurse or a social worker would work with a consulting psychiatrist to integrate care.

Dr. Thomas Insel, former director of the National Institute of Mental Health, emphasized the importance of integrating mental and physical health care, "Put simply, the mental health problem is medical, but the solutions are not just medical—they are social, environmental, and political. We not only need better access to medical treatments; we need to include people, place, and purpose as part of care. With comprehensive high-quality care—care that includes people, place, and purpose—people can heal" [50]. One-quarter of Americans currently require mental health care but less than one-half of those have access to that care [50]. Mental illness has never been given parity with physical illness. It is time.

A Biopsychosocial Illness Model

In Section 3 of our book, we labored to provide a unified medical model that may help explain long-COVID syndrome. However, we are not losing sight of the impact that psychological and social factors have on every long-COVID patient. Diseases are traditionally considered in terms of a strict biomedical model. Symptoms are expressed in relationship to altered body pathophysiology, which is determined by the severity of organ damage. Yet, we all know that two people with the same disease often have markedly different symptoms and outcome. These differences are related to psychological and social factors. Therefore, if we are to optimally understand and treat long COVID, we need to examine it through a biopsychosocial lens.

In 1977, Dr. George Engel wrote a classic study suggesting that a strictly biomedical approach in medicine was too narrow. He focused on the importance of taking into account each person's illness experience [51]. He predicted that these psychosocial factors could be measured and need to be considered alongside biomedical data, not as an afterthought.

As we have mentioned repeatedly, long COVID has laid bare this centuries old mind or body schism. On one side, scientists believe that long COVID is a unique, new pathophysiological disease, most likely immune in nature. Others think long COVID is primarily a psychological disorder, with severe stress at its core. This polarization is obsolete since there is overwhelming evidence that all mind–body interactions are bidirectional. For example, it is not surprising that there is a striking increase in depression, anxiety, and insomnia after COVID-19 infection. What is more surprising is that prior psychiatric illness increases the risk of getting COVID-19. In other words, mental ill health is both a risk factor for and a complication of COVID-19 [52].

We think of stress as a psychological factor, but it has far-reaching physical consequences. Psychosocial factors, including depression and stressful life events account for one-third of the risk for a heart attack [53]. Stress as well as physical and emotional trauma greatly impact every chronic medical disease [54]. Increased social support is associated with better outcomes in diseases as diverse as rheumatoid arthritis, phantom limb syndrome, spinal cord injury and multiple sclerosis. Physical and psychological abuse are risk factors for every chronic pain condition, including fibromyalgia, irritable bowel syndrome, chronic pelvic pain, and temporomandibular joint disorders. In adult veterans, combat exposure and PTSD correlate with chronic pain, another example of the impact of stress on a common medical symptom.

The lack of any physical abnormalities in long COVID, as in ME/CFS, fibromyalgia, POTS and countless other medical disorders, adds to the medical controversy regarding their acceptance as diseases. Conditions with such hidden disabilities are more common in women, and the medical profession has a long history of dismissing female illnesses as psychological. They get written off as hysterical, stressed out.

As we have discussed, the female predominance in long COVID should also be examined through a biomedical and a biopsychosocial lens. Autoimmune diseases are much more common in women. Women have a more potent immune response to infections and vaccines. Estrogen has effects on B- and T-cell reactivity, and the onset of autoimmune diseases varies with menopause. There are sex differences in pain sensitivity and the prevalence of every chronic pain disorder in women, including CFS/ME, fibromyalgia and migraine headaches, have increased [55].

Psychosocial factors also interact with these biologic differences between men and women. Women are encouraged to be more sensitive to emotional stimuli and in touch with their feelings. They also are more introspective. This makes women more prone to focus on their health, which taken to extremes is called catastrophizing. Catastrophizing, considered a purely psychological factor, is associated with alterations in brain structure and function. Greater levels of catastrophizing correlates with altered brain connectivity in areas such as the insula and amygdala in patients with fibromyalgia, migraine, and phantom limb syndrome [54].

A strict biomedical disease system promotes a paternalistic physician-patient relationship. The doctor is a technical adviser, the patient a passive recipient. A biopsychosocial model fosters a patient-physician partnership. It takes into account pre-existing illness vulnerability, be it genetic (biologic) or environmental (social). A patient's individual psychosocial factors, such as mood and sleep, and external support systems are evaluated with their impact on biomedical pathways.

We hope our hypothesis that symptoms such as exhaustion, cognitive, sleep, and mood disturbances are disorders of altered brain circuitry will promote greater acceptance of the bidirectional mind–body interactions. The biopsychological, patient-centric model has been best achieved in chronic conditions such as diabetes where team-care has improved glucose monitoring, medication adherence, weight control, exercise, foot care, and psychological well-being [56].

One long-COVID patient was asked by her friend, "I'm not being awful, but do you think a lot of it's in your mind? and I said 'no'. I was quite upset about that" [6]. A healthcare professional with long COVID noted how her experience with long COVID had changed her outlook, "It's easier to say 'This is in your head than to say I don't have the expertise to figure this out.' Before COVID, I never once said to a patient, 'There's something going on in your body, but I don't know what it is.' I used to see medicine as innovative and cutting-edge, but now it seems like it has barely scratched the surface. My view of medicine has been completely shattered. And I will never be able to unsee it" [57].

We all seek a biomedical disease explanation. This seems feasible if long COVID follows a "one microbe, one illness" rule. Almost no infections are so simplistic, so linear. Take the relationship of stress and infections. During the Asian flu epidemic of 1957, individuals with pre-existing PTSD, depression, and anxiety had a threefold greater incidence of getting infected than those without significant stress or psychological issues [58].

The endocrinologist/scientist Dr. Hans Selye alerted medicine to the impact of stress on our immune system, "Overwhelming stress can break down the body's protective mechanisms. It is for this reason that so many maladies tend to be rampant during wars and famine. If a microbe is around us all the time and yet causes no disease until we are exposed to stress, what is the cause of illness, the microbe or the stress? I think both are and equally so" [59].

Healthcare professionals who have been experiencing long COVID as patients themselves have noted that many of their ideas about medicine changed, "People with chronic disease need time to really open up and explain their symptoms. Because we work in a stressed system, we don't have the time or mental space for those diagnoses that don't have easy answers. I never would have done that before. I would have just been afraid of the whole thing and found it overwhelming. I think those who are transformed by having the illness will be different people—more reflective, more empathetic, and more understanding . . . and long COVID will cause a revolution in medical education" [57].

Meghan O'Rourke, who has suffered from many of the invisible, chronic illnesses discussed in our book, nicely summarized their relationship to long

COVID, “In this sense, PTLDS, autoimmunity, ME/CFS, and long COVID are diseases of our era, conditions that illuminate the need for a shift in medical thinking, from the model of the specific disease entity with a clear-cut solution to the messy reality shaped by both infection and genetics and our whole social history, a reality that no one yet fully understands. In the absence of certainty, medical science remains unsure what story to tell. Too often it turns away from patients rather than listening to the long and chaotic stories we tell” [3]. Hopefully, the global attention attached to long COVID, with its subsequent major research funding, will provide a new understanding and therapy for millions of people with unexplained, chronic illness.

References

1 Castanares-Zapatero D, Kohn L, Dauvrin M, et al. Long Covid: Pathophysiology-epidemiology and patient needs. Health Services Research (HSR) Brussels: Belgian Health Care Knowledge Centre (KCE). 2021. KCE Reports 344. D/2021/10.273/31.

2 Spruit, M.A., Singh, S.J., Garvey, C. et al. (2013). An official American thoracic society/European respiratory society statement: key concepts and advances in pulmonary rehabilitation. *Am. J. Respir. Crit. Care Med.* 188: e13–e64.

3 O’Rourke, M. (2022). Unlocking the mysteries of Long COVID. *The Atlantic* (8 March).

4 Nopp, S., Moik, F., Klok, F.A. et al. (2022). Outpatient pulmonary rehabilitation in patients with long COVID improves exercise capacity, functional status, dyspnea and quality of life. *Respiration* https://doi.org/10.1159/000522118.

5 Newman, M. (2021). Chronic fatigue syndrome and long covid: moving beyond the controversy. *BMJ* 373: n1559.

6 Kingstone, T., Taylor, A.K., O’Donnell, C.A. et al. (2020). Finding the ‘right’ GP: a qualitative study of the experiences of people with long-COVID. *BJGP Open* 4: 5.

7 Ellingson, L.D., Stegner, A.J., Schwabacher, I.J. et al. (2016). Exercise strengthens central nervous system modulation of pain in fibromyalgia. *Brain Sci.* 6 (1).

8 Fontes, E., Bortolotti, H., da Costa, K.G. et al. (2019). Modulation of cortical and subcortical brain areas at low and high exercise intensities. *Br. J. Sports Med.* 16: https://doi.org/10.1136/bjsports-2018-100295.

9 Estévez-López, F., Maestre-Cascales, C., Russell, D. et al. (2021). Effectiveness of exercise on fatigue and sleep quality in fibromyalgia: a systematic review and meta-analysis of randomized trials. *Arch. Phys. Med. Rehabil.* 102 (4): 752.

10 Singh, S.J., Barradell, A.C., Greening, N.J. et al. (2020). British thoracic society survey of rehabilitation to support recovery of the post-COVID-19 population. *BMJ Open* 10: e040213.

11 Sallis, R., Rohm Young, D., Tartof, S.Y. et al. Physical inactivity is associated with a higher risk for severe COVID-19 outcomes: a study in 48 440 adult patients. *BMJ* 55: https://doi.org/10.1136/bjsports-2021-104080.

12 Moyer, M.W. (2022). 'I had never felt worse': long covid sufferers are struggling with exercise. *New York Times*. (12 February).

13 Wang, C., Schmid, C., Rones, R. et al. (2010). A randomized trial of tai chi for fibromyalgia. *NEJM* 363: 743–754.

14 Blitshteyn, S. (2021). Is postural orthostatic tachycardia syndrome (POTS) a central nervous system disorder. *J. Neurol.* 7: 1.

15 Weintraub, P. (2021). The way out of brain fog. *The Atlantic* (9 April).

16 Lee, S.H., Shin, H.S., Park, H.Y. et al. (2019). Depression as a mediator of chronic fatigue and post-traumatic stress symptoms in Middle East respiratory syndrome survivors. *Psychiatry Investig.* 16: 59.

17 van Koulil, S., van Lankveld, W., Kraaimaat, F.W. et al. (2010). Tailored cognitive-behavioral therapy and exercise training for high-risk patients with fibromyalgia. *Arthritis Care Res.* 62: 1377.

18 Lazaridou, A., Kim, J., Cahalan, C.M. et al. (2017). Effects of cognitive-behavioral therapy (CBT) on brain connectivity supporting catastrophizing in fibromyalgia. *Clin. J. Pain* 33 (3): 215–221.

19 Fowler-Davis, S., Platts, K., Thelwell, M. et al. (2021). A mixed-methods systematic review of post-viral fatigue interventions: are there lessons for long covid? *PLoS One* 16: e0259533.

20 Nobile, B., Durand, M., Olié, E. et al. (2021). The anti-inflammatory effect of the tricyclic antidepressant clomipramine and its high penetration in the brain might be useful to prevent the psychiatric consequences of SARS-CoV-2 infection. *Front. Pharmacol.* 12: 615695.

21 Guinot, M., Maindet, C., Hodaj, H. et al. (2021). Effects of repetitive transcranial magnetic stimulation and multicomponent therapy in patients with fibromyalgia: a randomized controlled trial. *Arthritis Care Res.* 73: 449.

22 Belluck, P. (2021). Some long covid patients feel much better after getting the vaccine. *New York Times* (17 March).

23 Strain, W.D., Sherwood, O., Banerjee, A. et al. (2022). The impact of COVID vaccination on symptoms of Long COVID. *Vaccines.* 10: 652.

24 Al-Aly, Z., Bowe, B., and Xie, Y. (2022). Long COVID after breakthough SARS-CoV-2 infection. *Nat. Med.* https://doi.org/10.1038/s41591-022-01840-0.

25 NBC (2022). Nightly News Full Broadcast (February 27th). https://www.nbcnews.com/nightly-news-netcast/video/nightly-news-full-broadcast-february-27th-134182469769 (accessed 20 May 2022).

26 Smout, A. (2022). Quarter of UK employers cite long COVID as driving absences. *Reuters* (8 February).

27 Briggs, A. and Vassall, A. (2021). Focusing only on cases and deaths hides the pandemic's lasting health burden on people, societies and economies. *Nature* 593: 502–505.

28 Morris, A. (2021). Another struggle for long covid patients: disability benefits. *New York Times* (27 October).

29 US Department of Health and Human Services (2021). *Guidance on "Long COVID" as a Disability Under the ADA, Section 504, and Section 1557.*

30 Phillips, S. and Williams, M.A. (2021). Confronting our next national health disaster-long haul covid. *N. Engl. J. Med.* 385: 577–579.

31 Burke, M.J. and del Rio, C. (2021). Long COVID has exposed medicine's blind spot. *Lancet Infect.* 21 (8): 1062–1064.

32 Thomas, L. (1979). *The Medusa and the Snail.* New York: Viking Press.

33 Lockshin, M.D. (2017). *The Prince at the Ruined Tower: Time, Uncertainty, and Chronic Illness.* New York: Custom Databanks, Inc.

34 Cooney, E. (2021). From medical gaslighting to death threats, long covid tests patients. *STAT* (22 November).

35 Ladds, E., Rushforth, A., Wieringa, S. et al. (2020). Persistent symptoms after covid-19: qualitative study of 114 "long covid" patients and draft quality principles for services. *BMC Health Serv. Res.* 20: 1144.

36 Callard, F. and Perego, E. (2021). How and why patients made long covid. *Soc. Sci. Med.* 268: 113426.

37 Perego, E., Callard, F., Stras, L. et al. (2020). Why the patient-made term "long covid" is needed. *Wellcome Open Res.* 5: 224.

38 Alwan, N.A., Blair, J.B., Bogaert, D. et al. (2020). From doctors as patients: a manifesto for tackling persistent symptoms of covid-19. *BMJ* 370: m3565.

39 Michelen, M., Manoharan, L., Elkheir, N. et al. (2021). Characterizing long COVID: a living systematic review. BMJ glob. *Health* 6: e005427.

40 La SEMG elabora un protocolo de atención básica para los afectados de COVID-19 persistente (2020). Sociedad Española de Médicos Generales y de Familia. https://www.semg.es/index.php/noticias/item/594-noticia-20200929 (accessed 2021 Jan 28).

41 Association COVID Long France (2021). *Qui sommes nous?* https://www.apresj20.fr/qui-sommes-nous (accessed 2021 Jan 28).

42 Garner, P., (2021). If you have long Covid as I did, don't give up hope: Recovery is possible. *The Guardian* (10 June).

43 Roth, P.H. and Gadebusch-Bondio, M. (2022). The contested meaning of "long COVID"-patients, doctors, and the politics of subjective evidence. *Soc. Sci. Med.* 292: 114619.

44 Sellers, F.S. (2021). Could long covid unlock clues to chronic fatigue and other poorly understood conditions? *The Washington Post* (20 November).

45 Bate, P. and Robert, G. (2006). Experience-based design: from redesigning the system around the patient to co-designing services with the patient. *BMJ Qual. Saf.* 15: 307.

46 Gerteis, M., Edgman-Levitan, S., Daley, J., and Delbanco, T.L. (ed.) (1993). *Through the patient's Eyes: Understanding and Promoting Patient-Centered Care.* San Francisco, CA: Jossey-Bass.

47 Berwick, D.M. and Hackbarth, A.D. (2012). Eliminating waste in US health care. *JAMA* 307: 1513.

48 Baker, R., Freeman, G., Haggerty, J.L. et al. Primary medical care continuity and patient mortality: a systematic review. *Br. J. Gen. Pract.* https://doi.org/10.3399/bjgp20X712289.

49 Gawande, A. (2009). The cost conundrum. *The New Yorker* (2 June).

50 Insel, T.P. (2022). *Healing: Our Path from Mental Illness to Mental Health.* New York: Penguin Press.

51 Engel, G. (1977). The need for a new medical model: a challenge for biomedicine. *Science* 196: 129.

52 Taquet M, et al. Bidirectional associations between COVID-19 and psychiatric disorder: retrospective cohort studies of 62 354 COVID-19 cases in the USA. *Lancet* 2020;8(2):130–40https://doi.org/10.1016/S2215-0366(20)30462-4.

53 Yusuf, S., Hawken, S., Ounpuu, S. et al. (2004). INTERHEART study investigators. Effect of potentially modifiable risk factors associated with myocardial infarction in 52 countries (the INTERHEART study): case-control study. *Lancet* 364: 937.

54 Meints, S.M. and Edwards, R.R. (2018). Evaluating psychosocial contributions to chronic pain outcomes. *Prog. Neuropsychopharmacol. Biol. Psychiatry* 87: 168.

55 Fillingham, R.B. (2000). Sex, gender, and pain: women and men really are different. *Curr. Rev. Pain* 4: 24.

56 Segal, L., Leach, M.J., May, E., and Turnbull, C. (2013). Regional primary care team to deliver best-practice diabetes care: a needs-driven health workforce model reflecting a biopsychosocial construct of health. *Diabetes Care* 36: 1898–1907.

57 Yong, E. (2021). Even health-care workers with long COVID are being dismissed. *The Atlantic* (24 November).

58 Cluff, L.E., Canter, A., and Imboden, J.B. (1966). Asian influenza: infection, disease and psychological factors. *Arch. Intern. Med.* 117: 159.

59 Selye, H. (1946). The general adaptation syndrome and diseases of adaptation. *J. Clin. Endocrinol.* 6: 217.

10

The Way Forward: For Patients, Healthcare Providers, and Research

In this final chapter we will summarize key information and formulate guiding principles to move forward and achieve a better understanding and management of long COVID. This will include our recommendations to patients with long COVID followed by our suggestions for health care providers. We will then review our thoughts on the ongoing and future research in long COVID, bringing patients, clinicians, and researchers together in a unified effort.

For Patients

Do I have Long COVID?

Long COVID has become a catch-all term for a huge variety of symptoms that are the result of an infection with SARS-CoV-2. Since there is not an agreed-upon long-COVID definition or helpful diagnostic tests, patients are unsure if they have the condition or where to go for help in making a diagnosis of long COVID. Long COVID should be considered in any person with a confirmed or suspected SARS-CoV-2 infection who is experiencing new medical symptoms that are significantly interfering with their life. Most patients with SARS-CoV-2 infection make a complete recovery, although they may not feel back to normal for a few weeks or up to one month. This takes much longer in any patient who was hospitalized. We believe that persistent symptoms lasting at least three months after the initial infection is an appropriate time frame for long COVID.

Lingering symptoms in hospitalized patients are generally related to well-defined organ damage, what we have termed long-COVID disease; the subsequent medical course is similar to that of any severely ill patient. In most hospitalized patients, it takes weeks to fully recover from COVID-19. In contrast, what we have called long-COVID syndrome should be considered when multiple new

Unravelling Long COVID, First Edition. Don Goldenberg and Marc Dichter.

symptoms have persisted for at least three months after the initial infection and when these symptoms are not explained by the usual course of an illness. Such long-COVID syndrome symptoms do not necessarily correlate with the severity of the initial SARS-CoV-2 infection and may develop after mild or even asymptomatic cases.

Although there are many symptoms associated with long COVID, fatigue is almost always a primary problem. It is described as an intense lack of physical energy and stamina along with mental tiredness. This includes decreased mental clarity, focus, and understanding. The fatigue is aggravated by physical or mental exertion and is not improved by simple bedrest. This post-exertional malaise (PEM) is disproportionate to the activity or exertion and may last for prolonged time periods. Other common long-COVID symptoms include shortness of breath, muscle and chest pain, difficulty sleeping, mood disturbances, and palpitations.

As a patient, you should list your persistent symptoms and describe their severity. In the United Kingdom, patients who may have long COVID are invited to complete online screening questionnaires at 10 to 12 weeks after their initial infection. One such survey, The Newcastle Post-COVID Syndrome Follow up Screening Questionnaire (http://www.postcovidsyndromebsolnhs.k), asks the following:

Have you made a full recovery or are you still troubled by symptoms? If not,

1) Are you more breathless now than you were before your COVID illness?
2) Do you feel fatigued (worn out/lacking energy or zest) compared with how you were before your COVID illness?
3) Do you have a cough (different from any cough you may have had before COVID-19)?
4) Do you get any palpitations? (you feel your heart pounding or racing)
5) How is your physical strength? Do you feel so weak that it still limiting what you can do (more than you were pre your COVID illness)?
6) Do you have any myalgia (aching in your muscles)?
7) Do you have anosmia (a diminished sense of smell)?
8) Have you lost your sense of taste?
9) Is your sleep disturbed (more than it was pre-COVID)?
10) Have you had any nightmares or flashbacks?
11) Is your mood low/do you feel down in the dumps/lacking in motivation/no pleasure in anything?
12) Do you find yourself feeling anxious/worrying more than you used to?
13) Have you lost weight since your COVID illness?
14) Any other symptoms (list)

Similar surveys can be accessed by patients and clinicians to aid in screening for long-COVID syndrome. We encourage both patients and healthcare providers to use such surveys.

Where Do I Go for a Diagnosis?

There is no simple test to confirm the diagnosis of long COVID, and the diagnosis is made based on a patient's symptoms. Your primary care provider (PCP) is usually the clinician initially tasked with making a diagnosis. A skilled PCP should be able to triage multisystemic complaints and exclude other conditions that may mimic long COVID. Your PCP, especially if you have a long-standing relationship, knows your prior medical history as well as concurrent medical, social, and psychological issues. Your PCP is the clinician best suited to refer you to appropriate specialists or to a long-COVID clinic. This should not be a pop-up clinic but rather an established, multispecialty clinic, ideally connected to a major medical center. These specialized centers have the expertise to confirm a diagnosis of long COVID.

Answering the question whether your symptoms are explained by the initial SARS-CoV-2 infection is often not straight-forward. For example, if you had dyspnea with an abnormal chest X-ray during the initial infection, persistent breathing difficulty for weeks or even months would be expected. Your PCP might repeat the chest X-ray or order chest imaging and PFTs. Pulmonary specialists, cardiologists and neurologists are often consulted to exclude persistent lung, cardiac, or brain damage. It is reassuring that patients without signs of organ damage during, or soon after, the initial infection rarely develop such damage months later. In contrast, if you had a mild initial SARS-CoV-2 infection, but three months later you are constantly exhausted, cannot think straight, and have widespread muscle pain, an early referral to a multidisciplinary clinic with expertise in long COVID would be warranted.

Assessing Illness Severity

You and your PCP should jointly determine the severity of your current symptoms. This is essential for determining subsequent referral to other healthcare providers. For example, a patient whose symptoms are mild should be able to care for themselves and go to work, although they may have to reduce work hours or job responsibilities. Patients with mild symptoms generally will not require specialist evaluation or care. Most long-COVID patients with mild or moderate symptoms might benefit from a referral to a physical therapist, a respiratory therapist, a cognitive therapist, or a mental health professional.

In contrast, a person with severe symptoms is unable to care for themselves and has extreme difficulty getting about or even just leaving the house. Patients with severe symptoms will require multidisciplinary evaluation and care. Patients with moderate to severe symptoms should be evaluated with consultations from various sub-specialists or through a long-COVID clinic. As discussed in Chapter 8, most long-COVID clinics require referral from a PCP although some long-COVID clinics will see self-referred patients. Criteria for more immediate referral to a long-COVID clinic or medical specialty services might include persistent:

- unexplained chest pain, palpitations, or shortness of breath;
- uncontrolled pain;
- autonomic symptoms consistent with POTS;
- patients not improving or getting worse over time;
- major mental health symptoms; and
- continued visits to a hospital or emergency departments.

General Principles You Should Know

Basic principles in discussing long-COVID syndrome should follow the guidelines used for conditions such as CFS/ME and fibromyalgia. We have based many of our recommendations for discussing and managing long COVID on two publications: The NICE guidelines for the treatment of CFS/ME [1] and the Multidisciplinary NHS COVID-19 Service to Manage Post-COVID-19 Syndrome in the Community, from the Leeds Primary Care Services, Leeds Community Healthcare NHS Trust and Leeds Teaching Hospital NHS Trust [2]. In addition, we have inserted some general thoughts regarding the pathophysiology of long-COVID syndrome to help answer common patient concerns. Each patient with possible long COVID should be told the following:

- Long COVID is a real, complex, chronic medical condition that may affect multiple body systems.
- It is a sequel to having had a SARS-CoV-2 infection.
- The pathophysiology is not well understood.
- It is important to distinguish patients with well-characterized organ damage (long-COVID disease), from those with persistent, unexplained symptoms (long-COVID syndrome). Those patients with long-COVID disease typically follow the course of individuals with other severe illness.
- We believe that most of the long-COVID syndrome symptoms originate in the brain and may be characterized as a brain-localized, immune dysregulation. These are demonstrated by biological abnormalities, such as from brain imaging or abnormalities of the immune and autonomic nervous systems. Such abnormalities help to distinguish long-COVID patients from matched individuals who never had COVID-19 or do not have long-COVID symptoms.
- This brain-immune dysfunction may help to explain similar post-viral conditions and disorders such as CFS/ME and fibromyalgia.

- External factors, particularly those that impact stress, play an important role in symptom severity and are often modifiable.
- Long COVID impacts everyone differently.
- Long-COVID symptoms often change unpredictably in nature and severity over a day, weeks, or months.
- Patients often sense disbelief in their healthcare professionals, which may foster a general loss of faith in the medical profession.
- Patients may feel stigmatized by family, friends, and healthcare professionals.
- Long-COVID symptoms can be managed but currently there is no cure. Most patients have improved and many no longer have symptoms.

Who Do You Listen To?

As we have discussed, long COVID was the first illness named and characterized by patients through social media, demonstrating the power of patient voices and their shared illness experience. This has fostered greater patient involvement in long-COVID research than in any prior new medical condition. Social media made such involvement accessible and helped set up comprehensive patient data registries. It has also fostered web-based long-COVID support groups and chat rooms, which are essential sources of social support and provide a sense of community. One long-COVID patient said, "And the best support I got was from all the Facebook groups, believe it or not. That's where I found a lot of information, because everyone else was on a similar timeline to me. So, we were all going through the same symptoms, so I knew I wasn't going crazy" [3].

Patients, clinicians, and researchers often have different points of view and expectations and must work together to optimally triangulate these three disparate voices. For example, we don't know how best to incorporate patients' lived illness experience with more traditional clinical research. This is especially critical when comparing findings of research acquired from social media, which may not be representative of the general population. In general, there is a disconnect between patients' lived experience and advice from healthcare providers and official sites [4]. Current research programs, such as the RECOVER initiative, are designed around patient's experiences with the illness.

Long-COVID care should be based on shared decision-making and includes patient self-monitoring and self-management. One of the cornerstones of this self-monitoring involves the principle of pacing. Pacing, whether used to begin an exercise program or to help a patient return to work, involves the following:

- Establish an individual's activity level that minimizes symptoms.
- Alternate rest and activity.

- Maintain activity level for a significant period of time before attempting to increase it.
- Consider graded exercise steps.

Avoiding Misinformation and Disinformation

Most of us derive much of our medical information from the internet and there are currently more than 80 000 websites devoted to health issues [5]. However, the quality of that information is often poor, especially when it comes to controversial or poorly understood conditions, such as long COVID. A spokesperson for the US Department of Health and Human Services commented that, "Trying to get information from the internet is like drinking from a firehose, and you don't even know what the source of the water is" [6]. There is no quality filter when accessing the internet or buying a health-related book.

Dietary interventions, including various vitamins, supplements, and additives, are touted for every chronic medical condition and long COVID is no exception. Diets that are gluten-free, aspartame-free, and vegetarian as well as omega-3 supplements, Chinese herbal medicines, and vitamins C and D have been discussed in social media and uncontrolled medical reports as helpful for long COVID. Some of these products do have antiviral activity, including flavonoids and other antioxidants, vitamin C, and zinc. As mentioned in Chapter 8, there are registered clinical trials for long COVID involving the use of diets high in whole grains, polyphenol-rich vegetables, and omega-3 fatty acid-rich foods as well as a low histamine diet, with limited cheeses, fruit, seafood, and nuts. There are also NIH-registered trials on probiotics, niacin, omega-3, coenzyme Q10, and various plant roots in long COVID. We believe that the use of these interventions is best done under supervision with a qualified specialist.

Meghan O'Rourke, author of The Invisible Kingdom, described the various alternative healthcare providers and dietary interventions she tried hoping to improve her multiple, unexplained symptoms, "All I had to do, according to the 'integrative' practitioners, was muster the will power to change my life. That larger problem, according to many members of my Facebook group, was a susceptible system, thrown off by toxins, stress, lack of sleep, and gut problems caused by an inflammatory diet, 'bad' bacteria, and unidentified food sensitivities. I embarked on a diet that had many enthusiasts among the cyber sick. It was a version of the 'specific carbohydrate diet', which looks a lot like the so-called Paleo regimen: no gluten, no refined sugar, little dairy, lots of organic meat and vegetables" [7].

It can be difficult for patients to distinguish fact from fiction, particularly when they are desperate for help. One area of major discordance involves the use of specific laboratory tests to diagnose long COVID. The only current utility of extensive laboratory testing in long COVID is to exclude other medical conditions. Some patients and clinicians insist on obtaining every imaginable test whereas others worry that a lack of testing may be perceived as too passive. On March 12, 2022, NBC News featured a story on a company in California hyping its unique test for long COVID, claiming that the serum levels of 14 cytokines would provide an accurate diagnosis [8]. Based on these cytokine test results, patients are advised to consult with the company's affiliated clinicians who may prescribe a combination of an HIV medication, blood pressure medications, a statin, and the controversial anti-parasite drug, ivermectin. These same tests and treatments are also recommended for patients who believe they have developed long COVID after receiving a COVID vaccine, as well as to patients with CFS/ME and chronic Lyme disease. The total cost for the testing, consultations, and treatments runs into thousands of dollars, which are not covered by insurance. By February of 2022, more than 14 000 individuals with self-diagnosed long COVID had signed up for this diagnostic and treatment program, and the company director said that his protocol has worked in 85% of patients. One long-COVID patient praised this company on social media, "I am hugely thankful to them. For the first time in this nightmare, doctors are telling me that they have things to try that they think may help me" [9].

However, there have been no peer-reviewed studies to back up the claims regarding this company's diagnostic tests or treatment results. Dr. Benjamin Abramoff, Director of the University of Pennsylvania Long-Covid Clinic, said "There's so much mystery about what causes long COVID and where the pathology is, it can be an enormous relief when somebody says it's your cytokines, these are what's causing you to feel this way. A lot of people think what they want is a pill, but we don't have that pill, and we're not particularly close to having it" [8]. Dr. Kanao Otsu, lead immunologist at the Center for Post-Covid Care and Recovery at National Jewish Health in Denver, said, "You don't need a cytokine panel to tell you that you're still suffering from your COVID that you had nine months ago" [9].

Patients should always consult with their PCP before undergoing diagnostic tests or considering medical therapy. The US and UK now have medical websites that provide high quality and up-to-date information on long COVID, such as the RECOVER site (http://recovercovid.org). The UK has a post-COVID information packet online with practical suggestions for symptom management, including breathing exercises, pacing, relaxation techniques, diet, and exercises (http://www.homerton.nhs.uk).

There are significant risks associated with self-prescribing, especially in individuals trying multiple interventions simultaneously. Research is beginning to accumulate on the use and misuse of treatments being tried for long COVID, and a study in the United Kingdom, called the Therapies for Long-COVID (TLC) Study, is surveying self-management in 4000 non-hospitalized patients with long

COVID and 1000 matched controls [10]. This study uses a new electronic patient-reported outcome questionnaire designed for long-COVID patients, termed the Symptom Burden Questionnaire for Long COVID, SBQ-LQ.

Disability and Disability Insurance

Up to 60% of long-COVID patients are unable to return to their previous full-time employment. Despite President Biden's July 2021 promise, "To make sure Americans with long COVID who have a disability have access to the rights and resources that are due under the disability law," Long-COVID patients continue to struggle to prove they are incapacitated [11] The headlines of a *Washington Post* article from March 8 2022, noted this persistent problem, "COVID long-haulers face grueling fights for disability benefits" [11]. Although long COVID qualifies as a disability under the Americans with Disabilities Act, it is difficult to meet the disability threshold that insurers insist upon, particularly with the paucity of objective tests to document a person's difficulties. Dr. Monica Verduzco-Gutierrez, Chief of the Long-COVID clinic at the University of Texas Health Science Center at San Antonio, complained, "Almost every day I'm filling out disability paperwork, writing letters of appeal, and talking with people from disability companies. I would say some denials are unjustified" [11]. The disability denial statement that a long-COVID patient received claimed, "Her symptoms are out of proportion to the diagnostic and exam findings and are inconsistent with functional impairment." She appealed this decision but was very discouraged, "The process is so difficult you feel like giving up. But what then? You go back to work to a job you can't do?" [11]. The US Equal Employment Opportunity Commission received more that 4000 allegations of disability laws being violated related to COVID-19 during 2020 and 2021 [11].

Terri Rhodes, CEO of the Disability Management Employer Coalition, suggests that an open dialogue is important and recommended that long-COVID patients tell employers, "I need help doing my job; here's things I think I can do; here's what I can't do. I don't know how long it will last. I want to get back to work, and I need help to do it" [12]. Rehabilitation specialists should work closely with long-COVID patients to help get the patient back to a normal lifestyle as soon as possible. Dr. Benjamin Abramoff, Director of the University of Pennsylvania's Long-COVID Clinic, discussed the importance of job flexibility, including remote work, "It's not uncommon for returning to work to re-trigger symptoms that had been improving or resolved. Remote work allows them to conserve energy by not having to deal with the commute, walking into the building, [or] settling into the office. That way, the energy that they do have can be devoted to work" [13].

Patients also need to be aware of the evidence that leaving work for prolonged periods of time may lead to adverse outcomes. Up to 50% of patients with CFS/ME and fibromyalgia have applied for Social Security disability [14]. Women with fibromyalgia and CFS/ME who continued to work had better quality of life than those who were unemployed. Patients who managed to keep their jobs as long as possible have a better prognosis. Often, protracted sick leave and long periods of not-working are damaging to both physical and mental health. Long-COVID patients may need intensive support to return to work, requiring changes in job description and employer flexibility. Long COVID has forced society to better recognize the profound physical and psychological impact of persistent, chronic, poorly understood symptoms, as noted by the headlines of a June 6, 2022, *Washington Post* article, "How long covid could change the way we think about disability" [15]. In this article, Ken Thorpe, former US deputy assistant secretary for health policy, suggested that long COVID involves "...a different mix of people than what we've seen in the traditional disability population, puts a different and important face on whole problem of long-term care. Collectively, we can be more effective highlighting the policy issues" [15].

Biopsychological Perspectives to Remember

Every chronic illness is best examined from a biomedical, as well as a biopsychological, lens. SARS-CoV-2 is a new infectious disease and naturally people reasoned that long COVID was a novel condition involving unique biomedical pathways. Since long COVID was first described, the three disease mechanisms receiving the most attention have been persistent viral infection, inflammatory organ damage, and immune abnormalities. There has been little attention devoted to biopsychological factors in long COVID.

People know that stress has a major impact on heart disease, but stress is rarely mentioned as an important factor in long COVID. Long-COVID patients may assume that any focus on external factors, such as stress, means they are being tossed into a psychosomatic category, as has happened often with CFS/ME and fibromyalgia patients. Patients may worry "the doctor thinks it's all in my head." Long-COVID patients early in the pandemic heard the same skepticism that has plagued CFS/ME and fibromyalgia patients. In April 2021, a *Wall Street Journal* article entitled "The Dubious Origins of Long Covid," suggested that long COVID was "largely an invention of vocal patient activist groups, enabling patient denial of mental illness and a victory for pseudoscience" [16].

Societal and gender roles aggravate this mind-or-body dichotomy. Illnesses that are more common in women are often dismissed as psychosomatic. For example, a military veteran returning from combat with a diagnosis of post-traumatic stress disorder

(PTSD) faces much less scrutiny regarding their disability than a woman diagnosed with CFS/ME. Women with poorly understood chronic illness more often will be labeled as one of the worried-well. O'Rourke describes her difficulty in finding balance in such medical uncertainty, "You can't be deterred when you know something's wrong. But you've also got to be willing to ask how much is in your head—and whether an obsessive attention to your symptoms is going to lead you to better health. The chronically ill patient has to hold in mind two contradictory modes: insistence on the reality of her disease, and resistance to her own catastrophic fears" [7].

Long-COVID patients need to embrace a bidirectional mind–body approach to their management. A biopsychological illness model makes a long-COVID patient more likely to accept treatments as disparate as anti-inflammatory and immunosuppressive medications, antidepressants, and cognitive behavioral therapy. If we think of mental health more like physical health, we can breach the mind–body schism. A disorder of brain connectivity, as we have outlined in long COVID, can then be thought of as a brain arrythmia, analogous to a cardiac arrythmia. Dr. Thomas Insel, former director of the National Institute of Mental Health, said, "If we want the benefits and rigor of medical science, mental illnesses are medical illnesses, no different from diabetes or cancer or heart disease. We may not have identified a specific lesion or a diagnostic test, but mental illnesses are fundamentally brain disorders with a biology that involves the same kind of cellular and molecular changes found in other medical illnesses. People with mental illness should therefore be treated in the same health care facilities and covered by the same insurance" [17].

We have discussed the importance of psychosocial factors in every chronic illness. Stress, sleep, and mood disturbances have profound influences on heart disease, rheumatoid arthritis, and migraine headaches, not just poorly understood medical conditions. While we await a better understanding of the biologic pathways that underlie long COVID, we need to recognize and treat its biopsychological factors (Figure 10.1). Patients must understand the external and internal factors that contribute to their symptoms.

For Healthcare Providers

Making the Diagnosis

The diagnosis of long COVID will most often be made in primary care. Initially, PCPs expressed significant concern with making a long-COVID diagnosis because of a lack of clinical guidelines and the absence of unified, national guidelines [18]. During the past year, however, there was a much greater awareness of long COVID, and diagnostic guidelines have been published. A comprehensive case

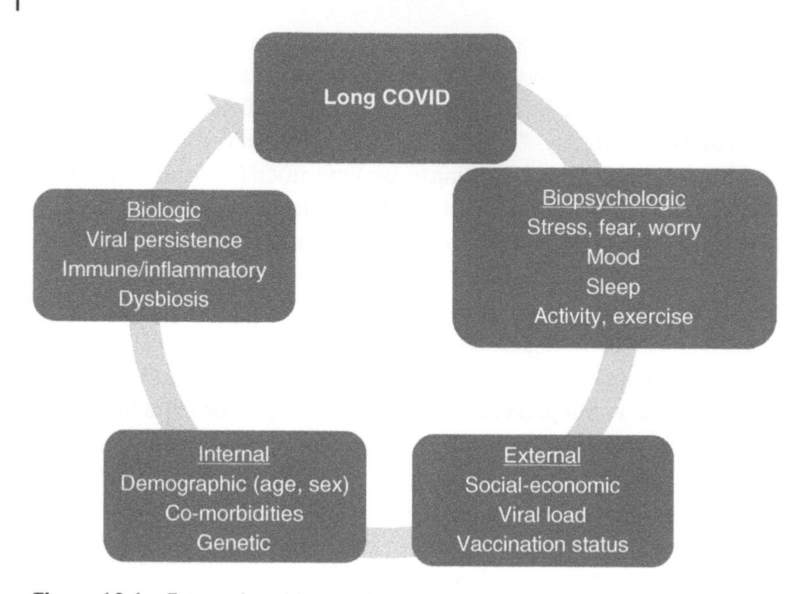

Figure 10.1 External and internal factors in long COVID.

report form for long COVID is available from the WHO that provides standardized questions about patient's symptoms, function, prior illness, and testing [19].

The long-COVID diagnosis is based on a detailed clinical history and physical examination. The physical examination and selected laboratory testing are required primarily to exclude other medical disorders. For example, joint swelling, significant muscle weakness, or neurologic abnormalities are not present in long-COVID syndrome. Screening tests to help exclude other conditions might include blood counts and chemistry, including muscle and liver enzymes, thyroid function, glucose, urinalysis, and acute phase reactants, such as erythrocyte sedimentation rate (ESR) and CRP. There is no current diagnostic utility for obtaining SARS-CoV-2 serology or immune or autoantibody testing, but this may change as research progresses. Surveys, such as the screening questionnaire mentioned above, can be useful and are available to insert into the patient record.

In one of the more comprehensive studies of diagnostic testing in long COVID, abnormal physical and laboratory findings were uncommon in long COVID and none of the extensive studies done by the investigators correlated with the presence of long COVID [20]. There was no correlation of long COVID with plasma levels of CRP, D-dimer, biomarkers of cardiac injury, such as troponin 1 or brain injury, including neurofilament light chain. There was also no correlation of long COVID with standard autoimmune tests, such as antinuclear or anticardiolipin antibodies and rheumatoid factor, or with selected immune/inflammatory markers, including numerous cytokines. The proportion of subjects with abnormal PFTs or transthoracic echocardiograms and 6-minute walking time did not differ in long-COVID patients compared to controls. In this study, the only risk factors for long COVID were the female sex and a self-reported history of prior anxiety disorder.

There is considerable debate about the specificity of long-COVID diagnostic criteria. We believe that in the initial evaluation for long COVID, a presumptive, clinical diagnosis should include the following: (i) multiple, new medical symptoms lasting for at least three months that (ii) began after a confirmed or presumptive SARS-CoV-2 infection, and (iii) cannot be explained by another disorder. Patients fulfilling these three criteria should be differentiated between those with organ damage, long-COVID disease, and those without evidence for organ damage, long-COVID syndrome.

Many investigators are concerned about the accuracy of proposed long-COVID diagnostic criteria, especially regarding potential bias in study design, such as recall and surveillance bias [21]. In research studies, it is important to apply uniform diagnostic criteria whenever possible. However, more flexibility with the long-COVID diagnosis is important in the clinic. For example, a positive, initial SARS-CoV-2 laboratory test to confirm the long-COVID diagnosis should not be insisted upon in a clinical setting.

A more universal issue concerns the aim of making a clinical diagnosis. As discussed earlier, the diagnosis of syndromes, such as CFS/ME, fibromyalgia, and irritable bowel syndrome, provide a framework for patients, clinicians, and scientists to better understand and treat these illnesses. We think of a syndrome diagnosis as a placeholder while awaiting more definitive evidence of disease. Until specific disease signs, tests, or pathology are identified, the diagnosis of a syndrome is determined by symptom description and severity. There is often no clear demarcation between people in the general population with symptoms, such as fatigue, widespread pain, headaches, or mood disturbances, from patients who are diagnosed with one of these common syndromes. For example, 10% to 15% of the general population have chronic, widespread pain but only 1% to 3% meet the diagnostic criteria for fibromyalgia. Similarly, 10–20% of the population report chronic, unexplained fatigue but less than 1% meet the criteria for CFS/ME. The diagnostic criteria allow us to distinguish individuals experiencing characteristic symptoms for extended periods of time and severe enough to adversely impact their daily activities. However, there is no evidence that the pathophysiology is different in people in the general population with chronic exhaustion or pain from those who are diagnosed with CFS/ME or fibromyalgia.

These caveats must be remembered when diagnosing long COVID. We need to consider the diagnosis as a proxy, with characteristic features that are not carved in stone. The diagnosis should include the three prime factors noted above, but flexibility is important. A symptom duration of three months is reasonable, but it is not a fixed time point. It is likely that more rigorous long-COVID diagnostic criteria will be devised in the near future. A study from the NIH RECOVER program used artificial intelligence and machine-learning, based on electronic health records, to aid in the diagnosis and sub-typing of long-COVID patients [22].

The next step is for the PCP to determine and document symptom severity. This can be accomplished with validated instruments, such as the RAND-36 summary scales, that measure physical and mental components of fatigue. Based on the symptom severity and the PCP's level of confidence, referral to selected specialists or a specific service, such as the long-COVID clinic, would be appropriate.

As discussed, long-COVID disease in hospitalized patients correlates with initial illness severity, such as patients with severe pulmonary disease. Long-COVID disease is more common in patients with co-morbid conditions and in older patients. Long-COVID syndrome is less-well-characterized, is more common in younger women, and does not correlate with initial symptom severity. Fatigue, dyspnea, cognitive, sleep, and mood disturbances are the most common symptoms in long COVID. Clinicians should recognize that long-COVID symptoms may change over time. For example, a loss of taste and/or smell and cough are very common during the first 3 to 6 months after the initial infection but quite unusual after 6 to 12 months and longer.

A clinician responsible for each patient's long-COVID care should be identified and assume patient follow up and continuity of care. Most often, the primary care team should take responsibility for the long-term care. As discussed in the previous chapter, conditions such as long COVID, CFS/ME, and fibromyalgia require that the patient and healthcare provider acknowledge illness uncertainty, but instead of saying "Well, we don't really know what's going on," should reassure the patient that "we don't know exactly what is going on, but we'll try to work it out."

We have discussed the importance of routine autonomic nervous system (ANS) testing in any patient suspected of having long COVID. If the PCP is not trained in basic ANS screening or if there is concern for POTS, a dysautonomia specialist should be consulted. The guidelines for pulmonary, cardiac, and neurologic testing continue to evolve and treatment will be directed by the appropriate specialist. In patients with long-COVID disease and evidence of organ damage, evaluation and treatment approaches follow those of any severe disease. In suspected long COVID, a chest CT scan is considered routine in patients with persistent dyspnea after three months and is often repeated in another three to six months. In patients with evidence of persistent pulmonary disease on imaging and PFTs, a course of corticosteroids might be recommended. The index of suspicion for subtle cardiac injury should be raised in older hospitalized patients. In a large US Veterans Administration Hospital study of 154 000 patients with COVID-19 and 5.6 million controls, there was a significantly greater incidence in a wide range of cardiovascular abnormalities up to one year after COVID-19 [23]. In this study the average age was 61 years and 89% were male. This study reminds us that the long-term cardiac sequelae following even mild or moderate SARS-CoV-2 infection may still be substantial. Further cardiopulmonary exercise testing, echocardiography, and cardiac MRI might be appropriate as determined by the coordinating pulmonologist and cardiologist.

Most patients should undergo a comprehensive neurocognitive evaluation, and a careful evaluation for mood and sleep disturbances. If abnormalities in neurocognitive function are revealed, these tests should be repeated at appropriate intervals, under the supervision of cognitive neurologists, to determine if the patient's decline is progressive, static, or improving. Each of these outcomes may require different treatment strategies for the patient and family. In Chapter 8, we reviewed the tests commonly used to evaluate patients with long COVID, and these tests should be validated in large patient populations. Some patients will require further endocrine, dermatologic, rheumatologic, and/or gastrointestinal evaluation.

Specialty Care

Long-COVID specialty care has three primary aims: (i) to confirm the diagnosis, (ii) to coordinate multidisciplinary care, and (iii) to evaluate specific therapies. This may be carried out within existing primary and specialty care networks or with newly established long-COVID clinics. We encourage in-depth, face-to-face consultations, but these may be supplemented with virtual care. If there is a long wait time for evaluation at a long-COVID clinic, the PCP should refer patients within their care system to an appropriate specialist, most often a pulmonologist, neurologist, and rehabilitation specialist. Community outreach rehabilitation is underused in the management of long COVID.

General principles helpful in other chronic illnesses are the cornerstone of current long-COVID therapy. A fatigue management strategy based on the four "P"s of planning, pacing, prioritizing, and positioning, was suggested [24]. Despite the controversy regarding exercise recommendations in long COVID, most long-COVID clinics do initiate a cautious structured and supervised exercise program. This is done in increments, starting with walking, breathing exercises, gentle stretching, and light muscle strengthening. If tolerated, gradual graduated aerobic training can be initiated, while carefully monitoring symptoms and vital signs, including pulse oximetry.

Cognitive behavior therapy (CBT) should be offered to most long-COVID patients. This includes self-management strategies to improve the person's functioning and quality of life. The RECOVER study includes a multicenter randomized, controlled trial of CBT in long COVID, targeting severe fatigue, using a CBT protocol called Fit after COVID, which was tested in CFS/ME patients [25]. The factors addressed in this CBT program include disrupted sleep–wake patterns, dysfunctional beliefs about fatigue, low or unevenly distributed activity levels, perceived low social support, problems with processing the acute phase of COVID-19, fears and worries regarding COVID-19, and poor coping with pain. The program can be implemented in-person or virtually.

Specialty care and long-COVID clinics should participate in clinical trials to evaluate specific therapies. For example, the HEAL-COVID study sponsored by Cambridge University, UK, and the National Health Service Foundation Trust is a three-year controlled, prospective study to determine potential long-range benefit of using a statin and blood thinner in patients with long COVID (http://www.heal-covid.net). Many patients are already taking a range of medications and supplements before reaching a specialist, and in these cases, it is important to discuss this with the patient. Some of the medications and supplements commonly used in CFS/ME and fibromyalgia centers that are being tried for long-COVID patients include gabapentin or pregabalin, duloxetine or milnacipran, naltrexone, antihistamines, and medications used to treat autonomic dysfunction. We believe that, whenever possible, long-COVID therapy should be initiated in a specialty center and evaluated as part of a collaborative, controlled, clinical trial.

Every healthcare provider must familiarize themselves with long COVID. If just 5% of patients have symptoms for more than two months following COVID-19, then about 7 million people in the United States are currently suffering from long COVID. However, recent reports estimate that nearly 20 million US residents and 7 million UK residents have evidence of long COVID. If those numbers are correct, every healthcare professional needs to participate in the evaluation and care of long-COVID patients. PCPs in the United Kingdom noted that the adequate care of long-COVID patients requires strong, efficient links to appropriate specialty referrals and the wider availability of non-pharmacological chronic disease therapies, such as exercise, CBT, and telemedicine [18].

Many long-COVID patients have very limited access to health services, often related to lack of healthcare coverage for costly tests, such as cardiac or brain imaging. Clinicians must help patients obtain such coverage and acknowledge the impact of long COVID on work and its potential for disability. There is a need for a comprehensive and uniform way to assess disability in long COVID. The Long COVID Episodic Disability Questionnaire was developed in the United States, United Kingdom, Canada, and Ireland to measure the presence and severity of disability [26].

The current wait-times of weeks to months to be seen at long-COVID clinics is unacceptable. By July 2022, there were 89 long-COVID clinics in England but only one in Northern Ireland and none in Wales and Scotland. Of the estimated 2 million UK residents struggling with long COVID, only about 4500 had been evaluated at a UK long-COVID clinic [27]. Helen Donovan, a leader of the Royal College of Nursing, said, “With over 2 million sufferers there aren’t enough specialist services to meet the growing demand, and the help patients get varies hugely across the country. Of the 2 million people self-reporting long COVID, only a fraction are aware of, or accessing, the treatment available" [27].

There is also inequitable access to these clinics, with middle-class white females predominating most long-COVID clinic populations. Dr. Monica Verduzco-Gutierrez, a San Antonio rehabilitation physician leading a national effort to promote equity in long-COVID care said, "We know certain communities are hit harder by COVID — marginalized communities. . . Black, Hispanic, Native American. This is not what a lot of the long-COVID clinics are seeing for sure." [28]. At the University of Washington's Long-COVID clinic, 83% of the patients are white, compared to 66% in the general community [28].

As a model for long-COVID clinics, the University College London Hospital Post-COVID-19 Service described their initial experience evaluating 1325 patients over 12 months, between April 2020 and April 2021 [29]. Patients who were hospitalized and needed pulmonary support or had lung imaging abnormalities were referred for an initial triage at six weeks, and all other patients with possible long COVID were referred to the long-COVID clinic after 12 weeks of persistent symptoms. At the clinic, each patient is first triaged by a physician and physical therapist and then scheduled for further appointments, if deemed appropriate. Approximately one-half of these patients receive further specialty evaluation and management that is generally divided into physical rehabilitation, respiratory physiotherapy, fatigue management, vocational support, and psychological support. The distribution of hospitalized versus non-hospitalized patients was equal, but the age, gender, symptom duration, most common symptoms, and ethnicity differed between hospitalized and non-hospitalized patients (Table 10.1) [29]. The significantly younger age of patients, greater number of female patients, longer duration of symptoms, and lower ethnic minority patient status are consistent with our description of long-COVID syndrome, in contrast to long-COVID disease (Table 10.1).

In this long-COVID clinic, functional impairment was prominent in all patients but more so in non-hospitalized patients, despite less organ damage and lower predicted mortality risk. These investigators confirmed the frequently observed lack of correlation between dyspnea and oxygen desaturation or cardiac and pulmonary imaging abnormalities in non-hospitalized long-COVID patients.

Research, Patients, and the Public

Basic Scientific Issues

We believe that brain-restricted autoimmunity is the mechanism most likely responsible for long-COVID syndrome. Much of the scientific data supporting this broad hypothesis is emerging as we write this section. In Chapter 6, we

Table 10.1 Characteristics of patients at University College London Long-COVID Clinic[a].

Characteristics	Hospitalized	Non-hospitalized
Median age (years)	58	45
Female (%)	43	68
Ethnic minority	53	31
Symptoms > 6 months (%)	39	50
Median number of symptoms	1	3
Shortness of breath (%)	39	60
Fatigue (%)	34	63
Chest pain (%)	14	31
Myalgia (%)	10	30
Headache (%)	7	29
Brain fog (%)	6	24
Sleep disturbances (%)	5	16
Loss of smell (%)	5	14
Postural symptoms (%)	3	13
Function (% best health)	80	60
Dyspnea scale score	35	57
PTSD score	28	35

[a] Adapted from [29].

reviewed the evidence for altered brain homeostasis in long-COVID syndrome, and, in Chapter 7, we discussed our theory that long-COVID syndrome is a localized neuroimmune disorder. Two other research hypotheses have linked long COVID to alterations of the microbiome and have suggested that long COVID is related to persistent SARS-CoV-2 virus or virus fragments sequestered in brain cells. These three research directions are not mutually exclusive but rather represent potential interacting processes in the development of long-COVID syndrome. The scientific issues related to these hypotheses are of special interest to researchers, and we have chosen to review this in depth in the Appendix.

As is the case in so much of modern science, the more we begin to understand complex mechanisms, the more questions we realize remain unresolved and in need of deeper understanding. There is clearly some hubris in this basic science section, given the breadth and depth of the biomedical community devoted to similar goals. Laboratories all over the world are turning their attentions to long COVID and new insights will appear in the near future. Our view is that the mechanisms underlying long COVID lie in the interface of three distinct disciplines: neuroscience, immunology, and virology. We believe that including this basic science section in a book intended for patients and their caregivers affirms the importance of patient-centered research, as discussed next.

Bringing Long-COVID Patients and Researchers Together

While scientists are working out the basic pathophysiologic mechanisms of long COVID, we must find ways to better incorporate patients in clinical research. Until recently, medical research has largely ignored patient descriptions and experiences of their symptoms. Symptoms, such as fatigue or joint pain, are subjective and considered unreliable measures of disease. In contrast, signs, such as joint swelling and inflammation, are objective disease parameters. Signs can be observed and measured by a qualified medical professional. Dr. Abigail Dumes, a medical and cultural anthropologist at the University of Michigan, noted, "When it comes to making a diagnosis, signs trump symptoms. This enduring hierarchy can be traced to the late 18th and early 19th centuries in the United States and Europe, when physicians who had relied on external symptoms for diagnosis shifted to a focus on internal anatomy and pathology by using technologies like microscopes. The French philosopher Michel Foucault observed that during that time, medicine transitioned from a practice in which the physician asked, 'What's the matter with you?' to a practice in which the physician asked, 'Where does it hurt?' [30]."

Signs, rather than symptoms, are the bedrock of clinical and basic research. In testing a new medication to treat rheumatoid arthritis, we have relied on physician-measured decreases in joint swelling, on improvement in X-ray joint damage, and in laboratory markers of inflammation. We have paid much less attention to patient self-evaluation of pain, fatigue, or quality of life. That started to change during the HIV epidemic, largely the result of patient activism. During the past 20 years, most clinical studies, such as in diseases like rheumatoid arthritis and MS, factor in patient opinion.

In poorly understood chronic illnesses without defined disease markers, patient experience is especially crucial to understanding and treating the condition. This is recognized in conditions such as CFS/ME and fibromyalgia but the reliance on subjective symptoms is often criticized as a research shortcoming. Three things have brought patient involvement in long-COVID research to the forefront: (i) the global attention on long COVID, (ii) the undeniable evidence that long COVID was a result of this unique infection, and (iii) the evolution of patient advocacy and research.

Patients named long COVID and have been intimately involved in its clinical research since it was first recognized as a medical disorder in early 2020. An early goal of long-COVID patient research was ". . .to capture and share a bigger picture of the experiences of patients suffering from COVID-19 with prolonged symptoms, using a data driven approach. The survey content and research analysis are 'patient-centric', conducted through participatory type research" [31]. Long-COVID patients, including physicians and many patients with research expertise, designed clinical surveys and often led the subsequent analysis and publication of these reports. Health care professionals are considered, "eloquent, informed 'expert patients' with knowledge of 'the system' and personal experience of its limitations. . . and advocate service improvements focused on the difficulty accessing investigations for persisting symptoms and the siloed, single-organ, and unintegrated nature of the system which led participants to propose holistic, 'one-stop shop' services integrating multiple specialties" [32].

The ZOE COVID app has been used by nearly five million individuals to provide in-depth patient self-reporting. In the ZOE COVID study, and other prospective studies of long COVID in the UK and the US, the characteristic symptoms are documented in real time and questions modified based on patients' experiences. Dr. Nisreen Alwan, a scientist–clinician at the University of Southamptom, advocated for patient involvement in long-COVID research, commenting, "I now feel like it's not legitimate to produce research which is not actually involving people with lived experience right from the start" [33].

A joint patient and clinician workshop at the University of Leicester, UK, in 2021 identified the most important research questions about long COVID [34]. The top 10 questions agreed upon were:

1) What are the underlying mechanisms of long COVID that drive symptoms and/or organ impairment?
2) What imaging techniques or scans may be able to detect and predict the development of organ problems or wider systemic issues?
3) What happens to the immune system throughout patients' recovery from COVID-19?
4) What can data at 6 and 12 months tell us about the long-term trajectory of illness?
5) What blood or other laboratory tests may be able to detect and predict the development of organ problems or wider systemic issues?
6) What is the impact of treatment(s) during the acute (initial) stage of COVID-19 on recovery?
7) What are the problems within the muscles associated with symptoms limiting activity/function/exercise? If so, what can be done to help?
8) What medications, dietary changes, supplements, rehabilitation and therapies aid recovery?
9) What can be done to support mental well-being during recovery?
10) What is the risk of future adverse health events (e.g. stroke, heart attack)?

Patient advocates are critical of the NIH RECOVER study's focus on newly acquired COVID-19 cases and that it is not studying patients who were infected more than two years ago. Karyn Bishof, founder and president of the COVID-19 Longhauler Advocacy Project, lamented, "Imagine that you've been profoundly ill for two years and told you're excluded from a project *literally called RECOVER* . . . that you helped advocate and secure funding for. Is anything planned for us, or are we excluded entirely?" [35]. Long-COVID patients, along with many physicians and scientists, expressed concern about the slow pace and lack of transparency of NIH-sponsored long-COVID research. Dr. Juan Wisnivesky, who co-directs Mount Sinai's Long-COVID Clinic, defended the NIH, stating, "Research has a slow pace, there is only so much you can push. I would say that the NIH is pushing very hard and very fast. If they were trying to do this faster, there might be more problems" [35]. However, even the NIH agrees that the enrollment of patients in the RECOVER study has been disappointing. The goal was to enroll 40000 long-COVID adults and children, but by March 18 only 1366 patients (3% of the goal) were enrolled [35]. Although many of the RECOVER investigators are involved in US long-COVID clinics, there is a disconnect between patients enrolling in RECOVER and accessing care at long-COVID clinics. For example, by March 2022, the Mount Sinai Long-COVID clinic recruited 1600 patients, more than were enrolled at all RECOVER sites combined [35]. Even in March 2022, many of the RECOVER centers listed on the NIH website were not accepting patients, more than six months after NIH funding began.

The NIH RECOVER study, and many studies from the United Kingdom, including the TLC study, are designed to address most of the concerns we have raised about previous clinical research in long COVID. This includes evaluating long-COVID patients and controls prospectively. As Dr. Shamil Haroon, a primary care specialist and co-lead investigator of the TLC study at the University if Birmingham, said, "It's important not just to look at the symptoms in isolation, but actually to look at what are they over and above what one would expect for a patient of a given description. Much of the literature does not have a control population, which is a real limitation. It makes it almost impossible to make any inferences about how the virus impacts an individual. What we need to do is compare long-COVID patients to others who are similar in every other respect, other than that they haven't had COVID-19. We know that patients report a very wide range of symptoms, but do those symptoms actually cluster together in ways that are due to different underlying, common disease pathways?" [36].

RECOVER is designed to accomplish this, and its design includes:

1) Patient stratification to include suspected, probable, and confirmed SARS-CoV-2 infection with matched control subjects. Long-COVID patients are compared to non-infected controls and infected subjects who did not have long COVID.

2) Prospective and retrospective design with a focus on new COVID-19 infections but a willingness to include those previously infected. It is a four-year longitudinal study.
3) A representative patient sample with an appropriate balance of inpatient, outpatient, and community subjects, with age, demographic, social, geographic factors. The target is 75% non-hospitalized subjects, including pregnant patients.
4) A detailed survey completed by all participants, either from a recorded review or in-person. A cohort of patients and controls will then undergo more detailed studies, such as advanced imaging and genetic, neuropsychological, and cardiovascular studies.

RECOVER tries to ensure equitable enrollment of long-COVID patients from ethnic and racial groups that were hit hardest by COVID. These groups are underrepresented in web-based surveys and in many long-COVID clinics. For example, Dr. Ingrid Bassett, a principal investigator at the Boston RECOVER site, noted, "The NIH enrollment goals are, for example, 27% of participants being Hispanic or Latinx, and 16% being African American. So these proportions are larger than these groups represent in the US population in general and that's really intentional, understanding those communities were very hard hit. In Boston, one of the ways we're doing that is through not only enrolling participants at the main academic center, which may be less accessible for certain people, but also going to our community health center partners, such as in Chelsea. We saw Chelsea as a very hard-hit area where there's a lot of engagement already with the community there around COVID-19. So this was a natural extension to say, 'This is something that's affected your community. And we're hoping that you can be part of the next steps for trying to treat and prevent this condition'" [37].

We need innovative ways to involve patients in every aspect of long-COVID research and integrate that with peer support and expert, updated information. Yale University long-COVID experts, Drs. Harlan Krumholz and Akiko Iwasaki, are working with patients and advocacy groups to develop new websites, such as Hugo Health platform; Kindred and the Patient-Led Research Collaborative has launched a new long-COVID survey, each designed to include patients in study design and data collection [38].

Treatment Trials

Often the goal of researchers differs from that of patients, whose primary goal is to find treatments that will improve their symptoms. Dr. Nabil Natafgi, an assistant professor in the department of health services policy at the University of South Carolina's Arnold School of Public Health noted, "What is relevant to policymakers and clinicians is not always what is most important for patients" [38]. The NIH RECOVER study was not designed to evaluate therapy, which may be its most glaring shortcoming, as noted by Dr. Lauren Stiles, a former long-COVID patient and an assistant professor of neurology at the State University of New York at Stony Brook, "This study is a slow-moving glacier. With a half-billion dollars, they could have run multiple clinical trials" [35]. In defense of the NIH RECOVER study, Dr. Walter Koroshetz, the director of the National Institute of Neurological Disorders and Stroke and co-chair of RECOVER, said that smaller, clinical trials were likely to be inconclusive, "My worry has been that unless we take a broad but engineered approach to this problem, in the worst-case scenario, we run a lot of small studies that can never be validated, we learn nothing, and we have no treatments" [35].

Currently, the only way to prevent long COVID is to avoid a SARS-CoV-2 infection, so it makes sense that vaccination reduces the chances of long COVID and decreases its severity. In one study from Israel, at 4–10 months after COVID-19 infection the fully vaccinated subjects were 50% to 80% less likely to have long-COVID symptoms than the unvaccinated (Table 10.2) [39]. The fully vaccinated subjects had no more likelihood of long-COVID symptoms than subjects who were never infected. The protection from vaccination was not found in subjects that received only one vaccine and was largely confined to infected patients more than 60 years old. The RECOVER study, and similar studies, will be investigating the protective effects of vaccination in long COVID prospectively.

Most of the initial long-COVID treatment studies simply extended ongoing therapeutic trials for acute COVID to determine if the intervention might lessen the likelihood of getting long COVID or decrease its severity. The WHO-sponsored SOLIDARITY study designed to evaluate antiviral and immune modulating drugs

Table 10.2 Long-COVID symptoms in unvaccinated versus vaccinated.

Percent of patients with each long-COVID symptom		
Long-COVID symptom	**Unvaccinated**	**Two vaccine doses**
Fatigue	26	11
Headache	22	14
Myalgias	11	6
Loss of concentration	10	4
Sleep disturbance	9	5
Shortness of breath	8	5

Source: Adapted from [39].

in acute COVID-19 is being reviewed for their potential impact on long COVID. The HEAL-COVID study was extended to determine if an anticoagulant or a lipid-lowering medication might be helpful in long COVID. The NIH-sponsored Accelerating COVID-19 Therapeutic Interventions and Vaccines (ACTIV-6) study that tracked antiviral drugs, monoclonal antibodies, and corticosteroids in acute COVID-19, is also being extended for patients with long COVID. The ACTIV-6 trial will test up to seven repurposed drugs in 13 500 non-hospitalized, mild to moderate COVID-19 patients. Although it was designed to evaluate the impact of these medications on the severity of the initial infection, it may provide data on their potential impact in long COVID.

It is difficult to design a long-COVID treatment study when the condition itself is not well-defined and there are no clear biomarkers to follow. Yet, that is no different than designing therapeutic trials for CFS/ME, fibromyalgia, chronic headaches, and depression. In such studies we have relied primarily on patient symptoms, and that is what we must rely on in studies of long-COVID therapies. The NIHR has funded a research study called STIMULATE (Symptoms, Inequalities, Trajectory, and Management: Understanding Long COVID to Address and Transform Existing Integrated Care Pathways). This study integrates clinical and imaging data with a treatment trial using existing drugs, including two antihistamines, loratadine and famotidine; an anticoagulant, rivaroxaban; and an anti-inflammatory medication used in gout, colchicine; to track efficacy in long-COVID symptoms.

Dr. Banerjee described the difficulty of drug trials in long COVID, "It's challenging, because we're going for a hazy target. People on the industry side are trying to figure it out too" [40]. Other novel drugs being investigated in long COVID include a medication developed to treat liver disease, a medication used to treat pulmonary fibrosis, and a drug that dissolves certain form of RNA that was tried for autoimmune diseases, such as SLE [40].

Long-COVID patients and clinicians expressed concern about the paucity of clinical treatment studies. Dr. Ryan Hurt, the head of Mayo Clinic's long-COVID study noted, "Federal agencies, foundations, and other funding entities should look at early, real-world pragmatic pilots for treatments soon. We have got to move a little quicker" [35]. Diana Berrent, founder of the long-COVID patient group Survivor Corps, said, "It's maddening that the NIH has decided to direct this massive amount of funding toward a four-year-long data collection project. . . . Their goals do not match the goals of people who are suffering" [35].

New, innovative ways to perform clinical trials have been advocated in order to speed up long-COVID treatment studies. The TOGETHER and RECOVERY studies in the United Kingdom used streamlined trials of multiple medicines and employed real-world evidence for drug efficacy. Rather than relying on the very costly, gold-standard, randomized, clinical trial, these trials are based on observational studies and very large databases. Sir Martin Landray, Professor of Medicine and Epidemiology at Oxford University, said, “One’s got to really think about how one can get better evidence at lower cost in order to get a better impact on public health. And that’s the motive. If trials continue to cost $1 billion or $2 billion each, we’re not going to get many new treatments for common diseases. And that’s not good value for patients and is not good value for those who pay for their care or deliver their care” [41].

A New Chronic Illness Strategy

There is hope that the global impact and recognition of long COVID, with appropriate government funding, will promote chronic illness research that had been sorely lacking in illnesses such as CFS/ME and fibromyalgia. The US$ 1.15 billion that the NIH has dedicated to studying long COVID is the first massive research investment in a chronic, poorly understood illness. For comparison, the NIH has allotted only US$ 35 million for CFS/ME research over the next five years. Despite this investment, many US COVID-19 experts have strongly rebuked the initial long-COVID research and offered suggestions to speed up and improve the situation [42]. They described research as “achingly slow” and called for much greater NIH funding of long-COVID therapeutic trials. As of February 2022, over 200 long-COVID clinical studies were listed on *ClinicalTrials.gov,* but only eight were funded by the NIH.

These public health experts suggested that the Biden Administration appoint a person to chair a long-COVID task force, creating a centralized agency to unify government response, accountability, and better engage the public [42]. Long-COVID clinics and government funding should be coordinated and integrated with RECOVER or similar research programs. For example, RECOVER should include programs, such as INSPIRE, a CDC-funded long-COVID study and ACTIV, the NIH’s platform to evaluate COVID-19 therapies. They suggested that within 90 days, the NIH launch a unified, open-access platform that integrates data from all existing long-COVID studies. A public database of all federally funded, long-COVID projects should be established and a nationwide messaging campaign to encourage patient enrollment in research should be launched.

On April 4, 2022, President Biden ordered more research and funding to be directed to long COVID, including a detailed report of available services and a new initiative, the “Health+ project,” to solicit feedback from long-COVID patients and expand US long-COVID clinics to include the Department of Veterans Affairs [43]. Dr. Harlan M. Krumholz, a professor of medicine at the Yale School of Medicine, noted, “It’s a landmark moment in long COVID. The White House [is] formally recognizing the magnitude of this health threat and formally committing to advancing knowledge in this area while simultaneously providing support for people who are suffering" [43]. Dr. Ezekiel Emanuel, the University of Pennsylvania bioethicist, cautioned that this announcement is a "good step in the right direction. . . but it needs to be bigger and faster. I get calls from people who are having fatigue or having shortness of breath upon minimal exertion. They really want some therapy. They want something" [43].

Approximately US$ 20 million has been earmarked for a project to study how health care systems can best organize and deliver care for people with long COVID, including using telehealth and expert consultation for primary care practices and advancing the development of multispecialty clinics to provide complex care. This will include primary care and specialist clinician education and support. Dr. David Putrino worried, "We are talking about $20 million to cover building out care coordination strategies for a mass disabling event that is affecting an estimated at least 2 percent of all Americans with a multisystem, multi-organ condition. This is complex care — $20 million doesn't get you very far." [43].

Dr. Amitava Banerjee, a lead investigator on long COVID at the University College, London, echoed our concern about preexisting attitudes toward chronic, unexplained illness, "Prevailing ways of thinking in science, healthcare, and policy haven't necessarily helped, either. For example, an outdated classification distinguishes diseases as 'organic' or 'functional'. Organic conditions, such as heart attacks, rheumatoid arthritis and bowel cancer are those that cause changes detectable by investigations such as blood tests or scans. Functional conditions, such as irritable bowel syndrome and chronic fatigue syndrome, do not necessarily cause changes detectable by tests, or the right test may not yet be available. Stigma and misconceptions arising from this classification may lead to functional conditions being overlooked, which is surely familiar to many with long COVID. This is a start, but we must do more. Long COVID, like all diseases, would benefit from public health and prevention perspectives, and 'integrated care' across specialties and disciplines. Siloing our thinking about disease and focusing on the short term has held us back, and patients have been left on their own for too long" [44].

This outdated mind-or-body illness model continues to plague our understanding of chronic illness like long-COVID syndrome. Long-COVID patients, including physician-patients, voice concern that, "clinical manifestations and research have overly focused on self-management, psychological support, and rehabilitation, resulting in the potential for 'watered-down' versions of long-COVID clinics that do not provide thorough physical assessment of patients" [45]. When patients with long COVID are offered cognitive therapy for brain fog, they may see this as psychological, rather than disease, management. Dr. James Jackson, director of behavioral health at Vanderbilt University, said, "If people begin to think of this as a brain injury, and not just as 'brain fog', they'll be more inclined to do what we do with people with brain injuries. Namely, we refer them to cognitive rehabilitation experts who can help them. With mild brain injuries many of these people get substantially better. But people don't think necessarily of cognitive rehab for 'brain fog'" [46].

The overlap of long-COVID syndrome with CFS/ME, other post-viral syndromes, fibromyalgia, and dysautonomia should provide important guidelines to understand and treat long COVID. Dr. Nisreen Alwan said, "I've so far resisted saying long COVID is ME/CFS, because I really think it is an umbrella term and there are multiple things happening in this long COVID umbrella. But for research, there should be a coalition" [47]. However, there is limited involvement of researchers from the fields of CFS/ME, fibromyalgia, and autonomic dysfunction with long-COVID research. A recent study called DecodeME, which was initiated to investigate genetic factors in 20 000 CFS/ME patients, is now enrolling patients who believe that their symptoms began with SARS-CoV-2 infection. The huge number of people affected by long COVID and its clear association with a viral infection has provided a natural experiment that may finally provide an understanding of conditions such as CFS/ME, fibromyalgia, and dysautonomia, as noted by Dr. Koroshetz, who said, "This is your opportunity. If you can't figure it out now, I think it's going to be really hard to ever figure this out" [35].

The association of dysautonomia with long COVID explains why many patients have normal cardiac and pulmonary testing yet report persistent exhaustion, dyspnea, chest tightness, palpitations, and lightheadedness. Borrowing from treatment programs in CFS/ME and fibromyalgia, long-COVID patients are being taught pacing, breathing exercises, and carefully, monitored exercise, starting in a sitting or recumbent position. Optimal rehabilitation of patients with long COVID, like CFS/ME and fibromyalgia, is based on balancing carefully incremental activity with adequate rest and recovery, and this needs to be acknowledged and covered by medical insurance.

Our healthcare system is not designed to treat chronic illness, especially conditions that are slow to mend. Dr. Lancelot Pinto, consultant pulmonologist at the P.D. Hinduja Hospital and Medical Research Center in Mumbai, noted, "Modern medicine is uncomfortable dealing with things where we don't have a quick fix. When there were no cures, patients were allowed to live out the natural history of the disease. For diseases that have a cure now, there is no leeway, it's presumed that if you are cured microbiologically, if the tests come back normal, you don't deserve any more rest. . . and that maybe the symptoms are imagined or psychological in some way" [48]. Jennifer Fagan, a previously very fit and healthy 48-year-old female, has suffered with persistent fatigue, shortness of breath, and palpitations since COVID-19 infection two years ago. She is working to rebuild her fitness levels but cautions that "Sometimes there's nothing to do but to slow down. And that's an OK thing" [49].

Summary

This book's title, *Unravelling Long COVID*, may strike some as too great of a reach. After all, long COVID is a new illness, not well characterized, and it is caused by a brand-new pathogen. Furthermore, long COVID is a heterogenous disorder, making its characterization difficult.

We believe that the lessons learned from similar medical conditions can be applied to long COVID to help elucidate much of its mystery. This unravelling begins with distinguishing two major long-COVID subsets, patients with well-characterized organ damage, what we term long-COVID disease, and patients with no organ damage or obvious disease biomarkers, what we term long-COVID syndrome. In patients with long-COVID disease, the persistent symptoms correlate with initial disease severity and subsequent organ damage. In Section 1, we reviewed the evidence that SARS-CoV-2 has resulted in significant chronic lung, heart, and central nervous system disease, with increased medical morbidity and mortality.

In contrast, long-COVID syndrome is not associated with well-defined organ damage. This is what most patients, health care providers and researchers mean when they discuss long COVID, persistent medical symptoms that are poorly understood yet can be disabling. This is the situation that patients and their health care providers find so confusing and controversial.

Such unexplained chronic illness is hardly a new phenomenon, often categorized as medicine's blind spot. That is why we have drawn on our experience in disorders such as CFS/ME and fibromyalgia, to help elucidate long-COVID syndrome. We have also compared other post-infectious illnesses to long COVID since they provide important clinical and research guidelines.

We believe that central nervous system dysfunction is the driving force behind long-COVID syndrome, and this is best explained by neuroimmune mechanisms. As these neuroimmune pathways become better elucidated, it is likely that more targeted therapy will be developed. In the meantime, long-COVID syndrome is best understood and managed with a comprehensive biopsychological illness model. External factors, particularly those that impact stress, play an important role in symptom severity, and these are often modifiable (Figure 10.1).

The overwhelming number of patients with long-COVID syndrome has forced patients, healthcare providers, researchers, and the government to finally pay adequate attention to the poorly understood, chronic illnesses that impact our minds and bodies. In the coming months we anticipate innovative discoveries and new research that will translate into better management of long COVID and similar poorly understood chronic illnesses.

References

1 National Institute for Health and Care Excellence. (2021). Myalgic encephalomyelitis (or encephalopathy)/chronic fatigue syndrome: diagnosis and management. *NICE guideline NG206.*

2 Parkin, A., Davison, J., Tarrant, R. et al. (2021). A multidisciplinary NHS COVID-19 service to manage post-COVID-19 syndrome in the community. *J. Prim. Care Community Health* 12, 21501327211010994. https://doi.org/10.1177/21501327211010994.

3 Burton, A., Aughterson, H., Fancourt, D. et al. (2022). Factors shaping the mental health and well-being of people experiencing persistent COVID-19 symptoms or 'long COVID.'. *BjPsych. Open* 8: e721.

4 Ladds, E., Rushforth, A., Wieringa, S. et al. (2020). Persistent symptoms after Covid-19: qualitative study of 114 "long Covid" patients and draft quality principles for services. *BMC Health Serv. Res.* 20: 1144.

5 Cline, R.W.J. and Haynes, K.M. (2001). Consumer health information seeking on the Internet: the state of the art. *Health Educ. Res.* 16: 671.

6 McLellan, F. (1999). 'Like hunger, like thirst': patients, journals, and the Internet. *Lancet* 352 (Supplement 2): 39.

7 O'Rourke, M. (2013). What's wrong with me? *The New Yorker* (19 August).

8 Edwards, E. (2022). Long Covid patients, in search of relief, turn to private company. *NBC News* (12 March).

9 Butler, K. (2022). Desperate patients are shelling out thousands for a long Covid cure: Is it real? *Mother Jones* (3 January).

10 Haroon S, Nirantharakumar K, Hughes SE, et al. (2021). Therapies for Long COVID in non-hospitalized individuals-from symptoms, patient-reported outcomes, and immunology to targeted therapies (The TLC Study): Study protocol. *medRxiv*. https://doi.org/10.1101/2021.12.20.21268098

11 Rowland, C. (2022). Covid long-haulers face grueling fights for disability benefits. *The Washington Post* (8 March).

12 Miller, K.L. (2022). How to deal with long covid at work. *The Washington Post* (17 March).

13 Lowenstein, F. (2022). How managers can support employees with Long COVID. *MIT Sloan Manag. Rev.* 14.

14 Walitt, B., Nahin, R.L., Katz, R.S. et al. (2015). The prevalence and characteristics of fibromyalgia in the 2012 National Health Interview Survey. *PLoS One* 10 (9): e0138024.

15 Sellers, F.S. (2022). How long covid could change the way we think about disability. *The Washington Post* (6 June).

16 Tuller, D. (2022). The medical establishment gaslights doctors, insisting long Covid is psychological. *Coda* (24 March).

17 Insel, T. (2022). *Healing*. New York: Penguin Publishing.

18 Brennan, A., Broughan, J.M., McCombe, G. et al. Enhancing the management of long COVID in general practice: a scoping review. *BJGP Open* https://doi.org/10.3399/BJGPO.2021.0178.

19 World Health Organization. (2021). Global COVID-19 Clinical Platform Case Report Form (CRF) for Post COVID condition (Post COVID-19 CRF).

20 Sneller, M.C., Liang, C.J., Marques, A.R. et al. (2022). A longitudinal study of COVID-19 sequelae and immunity: Baseline findings. *Ann Intern Med. Epub* https://doi.org/10.7326/M21-4905.

21 Wisk, L.E., Nichol, G., and Elmore, J.G. (2022). Toward unbiased evaluation of postacute sequelae of SARS-CoV-2 infection: challenges and solutions for the long haul ahead. *Ann. Intern. Med.* 175: 740–743.

22 Reese J, Blau H, Bergquist T, et al. (2022). Generalizable long COVID subtypes: Findings from the NIH N3C and RECOVER programs. *medRxiv*. https://doi.org/10.1101/2022.05.24.22275398

23 Xie, Y., Xu, E., Bowe, B. et al. (2022). Long-term cardiovascular outcomes of COVID-19. *Nat. Med.* https://doi.org/10.1038/s41591-022-01689-3.

24 Critical Care Forum, Royal College of Occupational Therapists Specialist Section – Trauma and Musculoskeletal Health (2022). Royal College of Occupational Therapists. *How to manage post-viral fatigue after COVID-19: Practical advice for people who have recovered at home.*

25 Kuut, T.A., Muller, F., Aldenkamp, A. et al. (2021). A randomized controlled trial testing the efficacy of Fit after COVID, a cognitive behavioral therapy targeting sever post-infectious fatigue following COVID-19 (ReCOVer): study protocol. *Trials* 22: 867.

26 O'Brien, K.K., Brown, D.A., Bergin, C. et al. (2022). Long COVID and episodic disability: advancing the conceptualisation, measurement and knowledge of episodic disability among people living with Long COVID – protocol for a mixed-methods study. *BMJ Open* 12: e060826. https://doi.org/10.1136/bmjopen-2022-060826.

27 Campbell, D., Geddes, L. (2022). Care for 2m Britons with long Covid 'woefully inadequate', say top nurses. *The Guardian.* (6 June).

28 Freyer, F.J. (2022). Groups hardest hit by COVID-19 appear least likely to get care for its lingering effects. Boston Globe (7 March).

29 Heightman, M., Prashar, J., Hillman, T.E. et al. (2021). Post-COVID-19 assessment in a specialist clinical service: a 12-month, single-centre, prospective study in 1325 individuals. *BMJ Open Resp. Res.* 8: e001041.

30 Dumes, A.A. (2022). What Long Covid shows us about the limits of medicine. *The New York Times* (17 March).

31 Assaf G, Davis H, McCorkell L, et al. (2020). Patient led research collaborative. Report: What does COVID-19 recovery actually look like? https://patientresearchcovid19.com/research/report-1/ (accessed 20 May 2022).

32 Ladds, E., Rushforth, A., Wieringa, S. et al. (2021). Developing services for long COVID: lessons from a study of wounded healers. *Clin. Med. (Lond).* 21: 59.

33 Gross, R.E., Boissoneault, L. (2022). Long covid could change the way researchers study chronic illness. *The Washington Post.* (17 March).

34 Houchen-Wolloff, L., Poinasarny, K., Holmes, K. et al. (2022). Joint patient and clinician priority setting to identify 10 research questions regarding the long-term sequelae of COVID-19. *Thorax* https://doi.org/10.1136/thoraxjnl-2021-218582.

35 Cohrs, R. (2022). 'A slow-moving glacier': NIH's sluggish and often opaque efforts to study long Covid draw patient, expert ire. *STAT* (29 March).

36 Mullard, A. (2021). Long COVID's long R&D agenda. *Nat. Rev.* 20: 329–331.

37 Cooney, E. (2022). 'We're all in this together': As long as Covid studies continue, researchers cast a wider net. *STAT* (7 March).

38 Sellers, F.S. (2022). How long covid is accelerating a revolution in medical research. *The Washington Post* (3 April).

39 Kuodi, P., Gorelik, Y., Zayyad, H., et al. (2022). Association between vaccination status and reported incidence of post-acute COVID-19 symptoms in Israel: a cross-sectional study of patients tested between March 2020 and November 2021. *medRxiv*. https://doi.org/10.1101/2022.01.05.22268800.

40 Rigby, J., Steenhuysen, J. (2022). Drugmakers, scientists begin the hunt for long COVID treatments. *Reuters* (25 March).

41 Herper, M. (2022). New Covid trial results may point toward better ways to study medicines. *STAT* (23 March).

42 Albarracin, D., Bedford, T., Bollyky, T., et al. (2022). The Rockefeller Foundation. *Getting to and sustaining the next normal: A roadmap for living with Covid.*

43 Diamond, D., Sellers, F.S. (2022). Biden announces long covid strategy as experts push for more. *The Washington Post* (5 April).

44 Banerjee, A. (2022). Opinion: I'm leading a long Covid trial-it's clear Britain underestimated its impact. *The Guardian* (12 January).

45 Gorna, R., MacDermott, N., Rayner, C. et al. (2021). Long COVID guidelines need to reflect lived experience. *The Lancet* 397: 455.

46 Flynn, H. (2022). Long COVID: Primate study reveals forms of 'brain injury'. *MedicalNewsToday*. (7 April).

47 Marshall, M. (2021). The four most urgent questions about long COVID. *Nature* 594: 168–170.

48 Chandrashekhar, V. (2022). Can this 19th century health practice help with long COVID? *The National Geographic.* (8 April).

49 Sheikh, K. (2022). How to improve heart health after Covid. *New York Times* (9 April).

Appendix A

Long-COVID Clinics in the US and Europe

Below is a list of long-COVID clinics in the USA as of February 1, 2022. From Survivor Corps Directory (survivorcorps.com). accessed February 1, 2022.

Alabama

University of Alabama at Birmingham COVID Respiratory Clinic

Alaska

Dignity Health COVID-19 and Chronic Illness Recovery and Reconditioning Program

Arizona

Banner Health Long COVID Treatment Program
Mountain Valley Regional Rehabilitation Hospital
Rehabilitation Hospital of Northern Arizona

Arkansas

Medical Associates of Northwest Arkansas Post COVID-19 Recovery Program

California

Los Angeles

Cedars-Sinai COVID-19 Recovery Program
Cedars-Sinai Post-COVID-19 Cardiology Clinic
Children's Hospital Los Angeles Long COVID Recovery Care
Keck Medicine of USC COVID Recovery Clinic

Others

Mission Heritage Medical Group Post COVID Clinic
Pomona Valley Hospital Medical Center (PVHMC) Post-COVID-19 Recovery Program
Sharp Allison deRose Rehabilitation Center COVID-19 Recovery Program
Scripps Health COVID Recovery Program

Unravelling Long COVID, First Edition. Don Goldenberg and Marc Dichter.

St. Jude Medical Center Post-COVID Rehabilitation Program
Stanford Post-Acute COVID-19 Syndrome (PACS) Clinic
UC Davis Health Post-COVID-19 Clinic Sacramento
UC San Diego Health Post-COVID Care Clinic
UCI Health COVID Recovery Service (Irvine)
UCSF OPTIMAL Clinic (San Francisco)

Colorado

COL UC Health - University of Colorado Hospital Post-Intensive Care Clinic
Lovelace Health System Post-COVID Recovery Services
National Jewish Health Center for Post-COVID-19 Care and Recovery

Connecticut

Gaylord Specialty Healthcare Post-COVID Outpatient Center
Hartford Healthcare COVID Recovery Center
Middlesex Health Post-COVID Rehabilitation
Post COVID-19 Recovery Program at Yale Winchester Chest Clinic
UConn Health Long-COVID Recovery Center
Yale New Haven Children's Post-COVID Comprehensive Care Program

Delaware

ChristianaCare Center for Virtual Health COVID Recovery Clinic

Idaho

St. Luke's Clinic – COVID Recovery
Rehabilitation Hospital of the Northwest

Illinois

Advocate Health Care Post-COVID Recovery Clinic
Edward-Elmhurst Health Post-COVID Neuro Care Clinic
Humboldt Park Health Post-COVID-19 Clinic
Northwestern Medicine Comprehensive COVID-19 Center
Northwestern Memorial Hospital Neuro COVID-19 Clinic
Rush University Medical Center Post-COVID Care Clinic
Shirley Ryan AbilityLab COVID Care Unit
University of Chicago Medicine Post COVID Recovery Clinic

Iowa

University of Iowa Health Care Post COVID-19 Clinic

Florida

Memorial Primary Care Long Haulers Clinic
University of Miami Health System COVID-19 Heart Program
Watson Clinic Post COVID-19 Clinic

Georgia

AbsoluteCARE Medical Center and Pharmacy Long COVID Clinic
Emory Executive Park Post-COVID Clinic
Grady Memorial Hospital Post-COVID Clinic
Piedmont Pulmonary COVID Recovery Clinic
Rehabilitation and Therapy Services at Augusta University Medical Center

Indiana

Parkview Health

Kansas

University of Kansas Health System Post-COVID-19 Care

Kentucky

Norton Healthcare COVID-19 Long-Term Care Clinic
Novak Center for Children's Health
UK HealthCare Pulmonary Rehabilitation Clinic

Louisiana

Our Lady of the Lake COVID Recovery Resources Program
Tulane Doctors Post COVID Care Clinic

Maryland, Washington DC

COVID-19 Recovery Center at the George Washington University
Crossings Healing and Wellness Post-COVID Care
Johns Hopkins Medicine Post Acute COVID Team (PACT)
Kennedy Krieger Institute – Pediatric Post COVID-19 Rehabilitation Clinic
MedStar Good Samaritan Hospital
MedStar Health COVID Recovery Program
MedStar National Rehabilitation Hospital
UM Rehabilitation and Orthopedic Institute COVID Recovery Rehabilitation Unit

Michigan

Michigan Medicine Physical Medicine and Rehabilitation
Pediatric Post-COVID Syndrome Clinic at Michigan Medicine C.S. Mott Children's Hospital
University of Michigan Post ICU Longitudinal Survivor Experience (U-M PULSE) Clinic

Massachusetts

Bay State Physical Therapy Post-Acute COVID Care
Beth Israel Deaconess Medical Center Critical Illness, COVID-19 Survivorship
Boston Children's Hospital Pediatric Long-Term Recovery
Brigham and Women's Hospital COVID Recovery Center

Emerson Hospital COVID-19 Recovery Program
Newton-Wellesley Hospital – Rehabilitation Services Post-COVID
Spaulding Rehabilitation Hospital Post COVID-19 Outpatient Rehabilitation Program

Minnesota

Mayo Clinic Long-COVID Clinic
Post-Acute COVID-19 Recovery at Hennepin Healthcare
University of Minnesota Adult Post-COVID Clinic

Mississippi

Recovery After COVID Clinic at Methodist Rehabilitation Center

Missouri

University of Missouri (MU) Mizzou Therapy Services' COVID Recovery Program
Washington University Complete Care and Recovery After COVID-19 (CARE) Clinic

Montana

Advanced Care Hospital of Montana

Nebraska

Children's Hospital and Medical Center Clinic for Pediatric Long Haulers

Nevada

Renown Health

New Hampshire

Dartmouth-Hitchcock Post-Acute COVID Syndrome Clinic

New Jersey

AtlantiCare Post-COVID-19 Long-Haul Clinic
CarePoint Health COVID Care Center
CentraState Post-COVID Syndrome Treatment Program
Deborah Heart and Lung Center Post-COVID Recovery Program
Franklin Cardiovascular Associates Post COVID Recovery Center
Hackensack Meridian *Health's* COVID Recovery Center
Hackensack Meridian Health's COVID Recovery Center Rehabilitation Program at JFK Johnson Rehabilitation Institute
Kessler Rehabilitation Center Recovery and Reconditioning Program
Monmouth Medical Center Southern Campus (MMCSC) Post-COVID Recovery Program
RWJBarnabas Health Post-COVID Comprehensive Assessment Recovery and Evaluation (CARE) Program
Saint Peter's University Hospital Post-COVID-19 Clinic
University Hospital Comprehensive COVID-19 Recovery Program

New Mexico

Rehabilitation Hospital of Southern New Mexico

New York

New York City Region

Bronx Care Health System Post-COVID Care Clinic
Columbia Primary Care Post COVID Care Program
Columbia University – Virtual Rehabilitation Program
Jamaica Hospital Medical Center Post-COVID Care Center
Maimonides Center for Post-COVID Care
Montefiore Department of Medicine CORE (COVID-19 Recovery) Clinic
Mount Sinai Center for Post-COVID Care
Mount Sinai Post-COVID Cardiac Clinic
Northwell Health Coronavirus Related Outpatient Work Navigators (CROWN) Program
Northwell Health COVID Ambulatory Resource Support (CARES) Program
NYU Langone Post-COVID Care Program
Pulmonary Wellness Online COVID Rehabilitation and Recovery Bootcamp
University Hospital of Brooklyn SUNY Downstate Post-COVID-19 Care Clinic
Weill Cornell Medicine Post-ICU Recovery Clinic

Others

Albany Medical College Post-COVID Care Clinic
Sunnyview Rehabilitation Hospital Long-Haul COVID Program

Nevada

Renown Health Post-COVID Care Program
UMC Southern Nevada Quick Care COVID Recovery Clinic

North Carolina

Cone Health Post-COVID Care Clinic
UNC Health COVID Recovery Clinic at the Center for Rehabilitation Care
WakeMed Rehabilitation Center COVID-19 Recovery Program

Oklahoma

Hillcrest Post-COVID Recovery Program
INTEGRIS Health's Post-COVID Recovery Program

Ohio

Cardiovascular Long-Term Recovery
Cleveland Clinic Post-COVID Cardiovascular Recovery Center
MetroHealth Post-COVID Clinic
Summa Health Post-COVID Clinic
The Ohio State University Wexner Medical Center Post-COVID Recovery Program

TriHealth PROS (Physical Rehabilitation Outpatient Specialists)
UC Health Post-COVID-19 Clinic
University Hospitals COVID Recovery Clinic

Oregon
Oregon Health and Sciences University Long COVID-19 Program

Pennsylvania
Allegheny Health Network (AHN) Post-COVID-19 Recovery Clinic
Main Line Health Post-COVID Recovery Program
Penn Medicine Lancaster General Health Post-COVID-19 Recovery and Rehabilitation
Penn Medicine Post-COVID-19 Neurological Care Clinic
Penn Medicine's Post-COVID Recovery Clinic
Temple Health Lung Center Post-COVID Recovery Clinic
University of Pittsburgh Medical Center (UPMC) Post-COVID Recovery Clinic

Texas
Baylor Medicine Post-COVID Care Clinic
Corpus Christi Rehabilitation Hospital
JPS Health Network COVID-19 Post-Recovery Clinic
Mesquite Rehabilitation Institute and Mesquite Specialty Hospital
Nexus Neurorecovery Center Post-COVID Program
South Texas Rehabilitation Hospital
Texas NeuroRehab Center COVID-19 Continuum Care Program
The Center for Post-COVID Recovery and Advanced Diagnostics
Trustpoint Rehabilitation Hospital of Lubbock
University of Texas Medical Department (UTMB) Post-COVID Recovery Clinic
UT Health Austin Post-COVID-19 Program
UT Health COVID-19 Center of Excellence
UT Southwestern Medical Center Department of Physical Medicine and Rehabilitation

Utah
Northern Utah Rehabilitation Hospital
The University of Utah Health Post-COVID Care Clinic
Utah Valley Specialty Hospital

Washington
UW Medicine Post-COVID-19 Rehabilitation and Recovery Clinic at Harborview Medical Center

West Virginia
Wheeling Hospital Post-COVID Outpatient Rehabilitation Program

Virginia

Centra Health Post-COVID Recovery Care Program
Hanover Rehabilitation Center
Inova Pos Sentara Heart Hospital Post-COVID Clinic
Sentara Heart Hospital Post COVID-19 Recovery and Rehabilitation Care Centers
Sheltering Arms Post-COVID-19 Rehabilitation Program
VCU Health Long COVID-19 Clinic

Vermont

UVM Health Network COVID Recovery Program

West Virginia

Wheeling Hospital Post-COVID Outpatient Rehabilitation Program

Wisconsin

Ascension Medical Group Post-Acute COVID Care
Bellin Health Long COVID Care
Froedert and Medical College of Wisconsin Post-COVID Multispecialty Clinic
Courage Kenny Rehabilitation Institute-Allina Health
Prevea Health COVID Recovery Clinic

Wyoming

Elkhorn Valley Rehabilitation Hospital

Long-COVID Clinics in the UK (as of February, 2022)

Barking, Havering and Redbridge University Hospitals NHS Trust, King George Hospital
Barnsley Healthcare Federation
Barts Health NHS Trust, Royal London Hospital and St Bartholomew's Hospital
Bath, Swindon and Wiltshire – Wiltshire Health and Care
Berkshire Healthcare NHS Foundation Trust
Birmingham Community Healthcare Trust
Bolton NHS Foundation Trust
Bradford Teaching Hospitals NHS Foundation Trust
Bristol, North Somerset and South Gloucestershire|Sirona Care and Health
Buckinghamshire Healthcare NHS Foundation Trust
Calderdale and Huddersfield NHS Foundation Trust
Cambridge University Hospitals NHS Foundation Trust
Chelsea and Westminster Hospital NHS Trust, Chelsea and Westminster Hospital
Community Surrey Health Surrey and Ashford and St Peter's Hospital
County Durham and Darlington NHS Foundation Trust

Doncaster and Bassetlaw Teaching Hospitals NHS Foundation Trust
Dudley Group NHS Foundation Trust
East Suffolk and North Essex NHS Foundation Trust
East Surrey and West Sussex – First Community Health and Care
East Sussex Healthcare NHS Trust
Essex Partnership University NHS Foundation Trust
Gloucestershire Health and Care NHS Foundation Trust
Gloucestershire Hospitals NHS Foundation Trust
Harrogate and District NHS Foundation Trust
Hertfordshire Community Service Health Care Trust
Homerton University Hospital Foundation Trust, Homerton University Hospital
Hull University Teaching Hospitals NHS Trust
Imperial College Healthcare NHS Trust, St Mary's Hospital
King's Health Partners (King's College Health Foundation Trust), King's College Hospital
Kings Health Partners (Guy's and St Thomas' NHS Foundation Trust), St Thomas' site
Lancashire and South Cumbria NHS Foundation Trust
Leeds Teaching Hospitals NHS Trust
Liverpool University Hospitals NHS Foundation Trust
London North West University Healthcare NHS Trust, Northwick Park Hospital
Manchester University NHS Foundation Trust (Manchester Royal Infirmary, Wythenshawe Hospital)
Mid and South Essex – Provide Community Interest Company
Milton Keynes University Hospital NHS Foundation Trust
Norfolk Community Health and Care Trust
North Care Alliance (Salford Royal NHS Foundation Trust, Royal Oldham Hospital)
North Cumbria Integrated Care NHS Foundation Trust
North Lincolnshire and Goole NHS Foundation Trust
North Manchester General Hospital
North Tees and Hartlepool NHS Foundation Trust
Northamptonshire Healthcare NHS Foundation Trust, Northampton General Hospital, Kettering
General Hospital
Northumbria Healthcare NHS Foundation Trust
Oxford Health NHS Foundation Trust
Oxford University Hospitals NHS Foundation Trust
Royal Berkshire NHS Foundation Trust – Berkshire Long COVID Integrated Service

Royal Surrey County Hospital
Sheffield Teaching Hospitals NHS Foundation Trust
Shropshire Community Health Trust
South Tees Hospitals NHS Foundation Trust
St George's NHS Trust, St George's Hospital
Stockport NHS Foundation Trust
Surrey Downs Health and Care Partnership
Sussex Community NHS Foundation Trust
Tameside and Glossop Integrated Trust
The Mid Yorkshire Hospital NHS Trust
The Newcastle upon Tyne Hospitals NHS Foundation Trust
The Rotherham NHS Foundation Trust
The Royal Wolverhampton NHS Trust
University College London Hospital Trust (UCLH)
University Hospitals Coventry and Warwickshire NHS Trust (UHCW)
University Hospitals Leicester NHS Trust and Provider Company Ltd. arm of the Leicester,
Leicestershire and Rutland Alliance
Virgin Care Services Limited and Frimley Health NHS Foundation Trust
Walsall Healthcare NHS Trust
Worcestershire Acute Hospitals NHS Trust
Wrightington Wigan and Leigh NHS Foundation Trust
York Teaching Hospital NHS Foundation Trust

Appendix B

Suggestions for Future Research Focused on the Cellular and Molecular Basis of Long-COVID Syndrome

Hypothesis 1: Brain-restricted Autoimmunity causes Long-COVID Syndrome

Background

Viruses are a major environmental factor that trigger autoimmune phenomena in genetically susceptible individuals. In an acute infection, viruses may initiate a hyper-reactive immune response, such as the cytokine storm of acute SARS-CoV-2 infections. In chronic disease, the relationship of the virus to autoimmunity is complex. In many chronic autoimmune diseases, such as systemic lupus erythematosus (SLE) or rheumatoid arthritis (RA), the prevailing theory is that a virus initiates the autoimmune disease, but the subsequent autoimmune reaction is self-sustaining. Viruses may remain dormant but, when activated, they trigger an immune response. Molecular mimicry is likely involved, whereby autoimmunity results from the fact that the virus is structurally similar to self-antigens. SLE is the prototypic autoimmune disease. The viruses most frequently associated with SLE are human endogenous retroviruses, Epstein–Barr virus (EBV), parvovirus B19, cytomegalovirus, and human immunodeficiency virus type 1, although a causal relationship of any single virus with SLE or RA has never been proven.

EBV is a potent activator of autoreactive B cells, which proliferate and generate antibodies with low affinity for host antigens that then attack cells and tissues. EBV also acts as a superantigen that stimulates numerous T cells to secrete large amounts of cytokines, which can be proinflammatory or immunoregulatory under different conditions. EBV is a latent infection in more than 80% of the US population, often present in B cells and pharyngeal tissues but in most people does not cause disease. However, EBV is associated with some severe diseases including Hodgkin's Disease, another form of lymphoma (Burkitt's lymphoma, mostly in Africa), mononucleosis, and multiple sclerosis (MS).

Unravelling Long COVID, First Edition. Don Goldenberg and Marc Dichter.

Interestingly, most researchers in the field consider MS to be an autoimmune disease confined to the brain, analogous to what might be occurring in long-COVID syndrome. The association of EBV with MS has been a major scientific conundrum for many decades. A recent study by the US military included over eight million active military with banked sera and known EBV status and identified 305 individuals who developed MS and matched them with 610 controls without MS [1]. Ten of the MS cases were initially negative for EBV (3.3%) but became positive before developing MS. The authors concluded "that MS risk is extremely low among individuals not infected with EBV, but it increases sharply in the same individuals following EBV infection."

In MS, there is a particular protein signal comprised of antibodies found in the cerebrospinal fluid (CSF), which indicates antibody production within the cranial cavity and not systemically, much like has been observed in long-COVID syndrome. These are called oligoclonal bands. Very recently, a group of MS researchers at Stanford University found that CSF lymphocytes from MS patients made antibodies directed at a specific EBV epitope that is also be a component of the myelin sheaths of central nervous system (CNS) neurons [2]. A further series of experiments indicated that this molecular mimicry may cause MS-like damage in experimental animals and produce an MS-like illness. Why this occurs in only a small number of people infected with EBV remains a mystery, as does the relapsing–remitting course of MS and its conversion to a progressive illness in some. These data provide a model for what may occur in patients who recovered from acute COVID-19 illness but develop mysterious symptoms weeks or months later.

Evidence in Long-COVID

Of particular interest, reactivated EBV is found circulating in the blood of individuals with COVID-19 infection [3] and in long-COVID patients [4]. This means that a SARS-CoV-2 infection can stimulate the conversion of a latent infection to a lytic infection whereby EBV actively replicates in its host cell, possibly via autoimmune mechanisms. One of the predisposing factors for long COVID was EBV reactivation during the acute COVID-19 illness, which is associated with the presence of certain autoantibodies, including those targeting interferons [5]. Several studies have already identified the presence of autoantibodies to neuronal antigens and multiple cytokines in the serum and CSF of long-COVID syndrome patients. The autoantibody and cytokine patterns were not necessarily the same in serum compared to CSF (Figure B.1) [6]. It has been suggested that brain-specific antibodies are produced by B cells residing within the blood brain barrier, which would account for the differences in CSF antibodies versus the serum [6].

Most of these studies did not differentiate between patients with long-COVID syndrome and those with initial organ damage. In addition, there is no solid

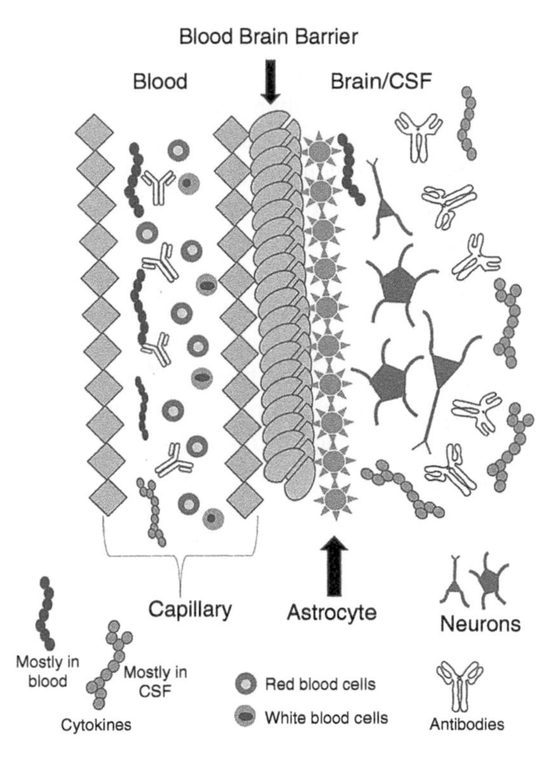

Figure B.1 Schematic illustration of brain capillary with red and white blood cells and antibodies and cytokines. Between the capillary and the cerebrospinal fluid (CSF) there is a relatively impermeable blood brain barrier (BBB). Within the CSF, there are astrocytes which help form the BBB, neurons within the brain, and antibodies and cytokines that are mostly different from those in the blood.

evidence that autoantibodies detected in the blood or CSF resulted in a traditional inflammatory reaction in the brain, as happens in several immune-related encephalitis conditions, such as anti-N-methyl-D-aspartate (NMDA) receptor encephalitis, discussed in Chapter 3. This is not a unique situation, however, as multiple paraneoplastic brain syndromes associated with specific anti-tumor antibodies do not produce a fulminate inflammatory response.

How Else Might SARS-CoV-2 Infection Influence the Immune Response Leading to Long COVID?

The SARS-CoV-2 virus can infect subsets of lymphocytes in a pathway that does not require ACE-2-receptors but does involve the viral spike protein. Viral RNA or antigens can be detected in CD4+ T cells from both blood and organs in COVID-19 patients. The virus induces apoptosis in these infected cells, accounting for the lymphopenia commonly seen in COVID-19 patients [7]. During acute infections, the patients with the lowest lymphocyte counts are often the most severely ill. Whether neurons or other brain cells that may become infected with the SARS-CoV-2 virus undergo a similar apoptosis is undocumented but could easily be missed if the cell loss is localized to small areas and ends early on in the initial infection, well before any post-mortem exams.

About 3% to 5% of CD4+ T cells are regulatory T cells, which are important in regulating the immune response to self-antigens. These cells are involved in a number of autoimmune diseases, including MS, type I diabetes, RA, and myasthenia gravis [8]. A significant loss of CD4+ lymphocytes may prevent suppression of the immune attack initiated by the systemic viral infection and possibly contribute to the cytokine storm that does significant damage during acute COVID-19.

Such a loss of CD4+ T cells could also reduce appropriate post-infection immune-regulation and thereby foster a persistent autoimmune attack that produces the long-COVID syndrome. At the present time, this is hypothetical, but hints of disordered CD4+ cell function in patients with long COVID are in the literature; this is a line of research that needs to be pursued [9]. It also should be noted that depending on ambient cytokines and other environmental factors, CD4+ T cells can foster the upregulation of the immune system within the context of complex immunological interactions. If most of the immune attack is confined to the brain and surrounding tissues, measurements of immune function only within serum will not necessarily identify etiologic factors in the brain and CSF and may, in fact, be misleading.

Therefore, measuring immune function in the brain and CSF and comparing it to that of the serum is important (Figure 1). Ideally, CSF biomarkers of long-COVID patients would be compared to healthy controls. Such studies have been done in other chronic illnesses. For example, a recent meta-analysis in patients with depression found that the CSF levels of interleukin-6 (IL-6), total protein, and cortisol, were higher among patients with unipolar depression, and levels of homovanillic acid, γ-aminobutyric acid, somatostatin, brain-derived neurotrophic factor, amyloid-β 40, and transthyretin were lower [10]. Blood and CSF during acute SARS-CoV-2 infection and acute neurological symptoms demonstrated a high frequency of autoantibodies targeting brain tissue [11].

Research that might support or refute the brain-specific autoimmune hypothesis:

1) Characterize the CSF and serum autoantibodies and cytokines in multiple patients with specific long-COVID phenotypes in detail.
   a) If the patterns are similar among patients with similar long-COVID phenotypes, it would support the hypothesis that the autoantibodies identified

are responsible for the symptoms. If, on the other hand, the autoantibody patterns appear to be unrelated to long-COVID phenotypes, it would raise serious questions as to their role(s) in pathogenesis.

2) Investigate autoantibody targets in the brain and how they correlate with symptoms, both in general and during exacerbations.
3) Testing for specific antibodies, such as to the receptor for the hypothalamic neuropeptide that stimulates arousal, orexin. Orexin has been correlated with patients taking longer to awaken after extubation and with sleep–wake disturbances and the downstream effects of chronic sleep deprivation.
4) Testing for IL-4, since it appears to be associated with cognitive abilities and there has been evidence that intranasal treatment with IL-4 improves cognitive disturbances after traumatic brain injury [12].
5) Testing for IL-6 (a pro-inflammatory cytokine), which is associated with hippocampal shrinkage in patients with depression [13].
6) Search for anti-idiotype antibodies that target the SARS-CoV-2 spike protein binding domain. Such antibodies would directly bind ACE-2 receptors.
   a) If such anti-idiotypes exist mimicking the SARS-CoV-2 spike protein, it is important to determine their mechanism of action on neurons or other brain cells that express the ACE-2 receptor, including if they activate the receptor, inhibit the normal activation of the receptor, or act to remove the ACE-2 receptor from the cell surface. Each of these has consequences to various brain networks. The enzymatic activity of ACE-2 is anti-inflammatory, and if it is inhibited or removed from the membrane surface, other pro-inflammatory effects would become apparent.
7) Determine the cellular mechanisms by which CSF autoantibodies and cytokines affect neuronal function in patients with long-COVID syndrome.
   a) Is there evidence for antibody-coated neurons (or other brain cells) in pathological studies? If yes, it may be possible to identify specific neurons in brain nuclei involved in specific functions related to long COVID, including dysfunctional autonomic sensory receptors (e.g. baroreceptors) other autonomic symptoms, disrupted sleep rhythms, memory problems, and others. If such neuronal regions are the targets for autoantibodies or anti-idiotype antibodies, there may be neurons missing, or injured, which could account for the circuit disruptions seen in long-COVID syndrome. This would likely require mammalian cell cultures, slice preparations, and/or animal studies.
8) Develop detailed descriptions of brain areas that demonstrate activated microglia and astrocytes.
   a) This can be initiated with positron-emission topography (PET) scans using microglial and astrocyte ligands. Such studies should also include all available post-mortem brain sections. Both activated microglia and activated astrocytes may secrete cytokines or other neuroactive chemicals that may

disrupt normal signaling pathways and critical networks without clear evidence of neuronal injury or loss. Both these cell types are ubiquitous in the brain and have the capacity to significantly alter brain function when perturbed. This is particularly important since some emerging therapies dampen microglial activation, which cannot be evaluated by routine blood or CSF studies.

b) The hypothesis of microglial activation in the habenula presented at the end of Chapter 7 suggests that specific imaging for activated microglia in this nucleus would be useful. Similarly, if post-mortem tissue becomes available from individuals with long COVID, it would be important to identify the specific brain regions involved in microglial activation to correlate these with the patients' symptoms?

9) Determine if specific neuronal types and non-neuronal brain cells are susceptible to SARS-CoV-2-induced apoptosis analogous to that of lymphocytes. This could be studied in tissue culture or mouse models of brain infection.

Potential therapeutic implications based on this research:

1) Antibodies that reverse IL-6-induced microglial activation.
2) Anti-intrathecal B-cell agents, such as those used in MS therapy.
3) Trials of NMDA receptor antagonists, such as ketamine, at doses used to treat therapy-resistant depressed patients. If such treatment reduces long-COVID symptoms, perhaps by dampening the microglial cytokine secretion, it would indicate the importance of further exploration.
4) Trials of drugs to suppress cytokines found in long-COVID syndrome patients.
5) Investigation of methods to restore immune-regulatory functions of CD4+ lymphocytes.

Hypotheses 2: SARS-CoV-2 Infection Alters the Microbiome, which in turn Initiates Long-COVID Syndrome

What is the Microbiome and How does it Affect Immunity and the Brain?

It is well known that the human body harbors many species of bacteria in the gastrointestinal (GI) tract and elsewhere in our healthy bodies. Most of the analyses focus on the bacteria colonizing the colon and lower GI tract, but the small intestine in the upper GI tract and even saliva also harbors microbiomes. There are so many different species in our colon that scientists have not identified many of them. Using new technologies, many bacteria of the microbiome can be identified by their genetics. In addition, patterns of how the different types of bacteria act on the immune system have been elucidated. Some microbiome patterns usually observed in healthy individuals can be distinguished from those in patients with a variety of illnesses. One microbiome pattern tends to be associated with a very active immune system and is called proinflammatory. Another pattern is thought to be more immunoregulatory, often dampening immune responses.

The microbiome has complicated interactions with the nervous system. For example, alterations of the microbiome, such as administration of some antibiotics, causes functional consequences in the brain. A similar problem occurs if the gut cells that transport essential amino acids from the small intestine to the blood are damaged and key chemicals cannot be actively transported into blood. A healthy microbiome supports those cells. Included in the essential amino acids the body cannot synthesize are tryptophan, phenylalanine, tyrosine, and histidine, all essential precursors for neurotransmitters.

The microbiome works as a two-way street in regard to the brain. Stress, both psychological and physical, causes changes in the microbiome. Changes in the microbiome can affect some behaviors including autism spectrum disorders, anxiety, depression, and chronic pain have been linked to altered microbiomes [14].

Many research studies designed to understand the normal role of the microbiome in healthy people rely on germ-free mice or rats. These are animals raised from birth in a germ-free environment and so, have no microbiome. Another model is to sterilize the GI tract with nonabsorbent antibiotics that wipe out, or greatly alters, the already existing microbiome. Germ-free mice have an exaggerated response to stress that can be reversed by adding a microbiome to these animals. Anxiety-like behavior, on the other hand, is reduced in the germ-free mice. Alterations in synaptic transmission in different brain regions have also been noted in germ-free mice. Many of these abnormalities can be eliminated by adding a normal microbiome. Interestingly, not all the features found in germ-free mice (a status originating at birth) are reproduced by eliminating the microbiome even early in life, indicating the importance of the developmental component of a normal microbiome. Dietary manipulation and fecal transplants can reduce the behavioral consequences of pro-inflammatory patterns in the mouse microbiome [15].

The microbiome interacts with the immune system, both systemically and in the brain. The development of microglia in the brain, the brain's main immune cell type, is influenced by the microbiome. The absence of microbiome in germ-free mice has a time- and sexually dimorphic impact both prenatally and postnatally [15]. In very young male mice, the absence of microglia was more profoundly perturbed than in females. On the other hand, antibiotic treatment of adults (to deplete the microbiome) more profoundly affects female mice. Such animal research may be relevant to understanding the effect of age and sex in patients with long COVID and determining the effects of altering the microbiome as a potential treatment strategy for long COVID. The microbiome in COVID-19 resembles that of patients with SLE.

The interactions between the microbiome and both the immune system and the brain are being better characterized [16]. This microbiome hypothesis is not meant to suggest that the microbiome itself is responsible for long-COVID syndrome but that its combined influence on the immune and nervous systems may affect the body's response to SARS-CoV-2 infections and long-term consequences after an acute infection. Importantly, this may offer possible novel therapeutic strategies for both components of COVID-19.

The Microbiome in COVID-19 Patients and in Individuals with Long-COVID Syndrome

Growing evidence for altered microbiome in the initial COVID-19 infection and in long COVID is developing. Patients admitted to the hospital with mild to severe symptoms of COVID-19 have microbiome patterns that differ from those of matched controls in the general population [17]. In patients who developed long COVID, these distinctive microbiome patterns persisted for at least six months after recovery from COVID-19. Specific microbiome differences included fewer immunoregulatory species, more pro-inflammatory species, and fewer butyrate-producing bacteria. These microbial patterns were seen regardless of initial COVID-19 severity, including in patients with mild disease. COVID-19 patients who did not develop long COVID had a microbiome pattern similar to uninfected control patients.

Antibiotic suppression of the microbiome is associated with memory deficits and reduced hippocampal neurogenesis, a result that can be reversed by a probiotic diet [18]. Gut butyrate-producing bacteria are important for multiple immune functions, such as reducing the microglial secretion of pro-inflammatory cytokines. Butyrate acetate and propionate, short-chain fatty acids (SCFAs), are presumed to play a key role in neuro-immuno-endocrine regulation as well as neurotransmitter synthesis and regulation. Patients with long COVID continued to have atypical microbiome patterns consistent with increased inflammatory species. At six months after COVID-19, there was a correlation of long-COVID symptoms and the microbiome composition. Those patients with an abundance of pathogens linked to opportunistic infections, including *Clostridium innocuum* and *Actinomyces naeslundii*, had more neuropsychiatric symptoms and fatigue. In addition, low microbiome butyrate was associated with decreased six-minute walk distances, possibly indicating a gut-to-lung interaction with the microbiome.

Similar changes in the microbiome have been found in patients with fibromyalgia [19] and CFS/ME [20]. Multiple studies show significant microbiome differences between fibromyalgia patients and controls, including bacterial species and serum metabolites influenced by the microbiome. In fibromyalgia, the microbial balance is toward pro-inflammatory species. Additional studies show differences between patients with CFS/ME and both their relatives and non-related control groups. The differences are in specific groups within the microbiome and point to specific pro-inflammatory changes as well as secondary microbiome-specific metabolic changes in the blood of the patients.

Research proposals that would support or reject the role of the microbiome in long-COVID syndrome:

1) More clearly identify harmful microbiome species from those that modulate immune function.
2) Identify specific species that play critical roles in overlapping long-COVID phenotypes. This would allow potential supplementation of the favorable species directly or elimination the harmful species.
3) Clinical trials that change the microbiome pattern to determine if such changes will reduce long-COVID symptoms. This could possibly identify useful treatments.
   a) The microbiome can change rapidly, sometimes over just a few days, with manipulation, including dietary changes. If the microbiome pattern changed from pro-inflammatory to immune regulatory with an improvement of long-COVID symptoms, it would be evidence for the direct effects of the microbiome on the prolonged illnesses and point to multiple therapeutic opportunities. This applies to any of the procedures described below:
   b) Probiotics and prebiotics that produce more regulatory microbiome patterns can be used. Diets like the "Mediterranean Diet" (fruits, grains, nuts, vegetables, fish) appear to be associated with this less inflammatory pattern, whereas paleo diets (high in meat, poultry, vegetables, seasonal fruits but no grains, nuts, and limited dairy products) may produce the opposite.
   c) Fecal transplants from individuals with favorable regulatory versus pro-inflammatory patterns into patients with long-COVID syndrome. Fecal transplants are used to treat patients with *C. difficile* infections and irritable bowel syndrome.
   d) Short-term administration of nonabsorbent antibiotics to eliminate or reduce the microbiome and permit recolonization with a better mix of organisms.
   e) Tailored microbiome changes by the supplementation of helpful species or removal of harmful ones with targeted bacteriophages.
   f) In patients with low levels of butyrate and the other SCFAs, a short-chain fatty acid replacement trial with the goal of reducing long-COVID symptoms.

g) Serum measurements of the essential amino acids that are precursors for neurotransmitters. If one or more of them are low, it could lead to many long-COVID brain-related symptoms. If such deficiencies are uncovered, the missing amino acids could be replaced as a therapeutic trial.
h) Characterize the relationship of microbiome patterns with long-COVID symptoms in rodent models of SARS-CoV-2 infections. Genetically modified mice develop phenotypes that very roughly correlate with human long-COVID symptoms. It would be useful to determine if these behaviors changed when the microbiomes were altered in specific ways. Research on the effect of SARS-CoV-2 on ACE-2-receptor expressing germ-free or conventional mice could provide important data about how critical the microbiome is in both the acute disease and its longer term consequences.

Potential therapeutic implications:

1) The microbiome is relatively easy to study and involves minimally invasive procedures, but requires a specialized laboratory setup.
2) The microbiome is relatively easy to change through several methods including special diets, the addition of specific foods, reduction with antibiotics, replenishment with other species, and through fecal transplants.
3) The effects of significant microbiome changes can be assessed with behavioral evaluations, serum and CSF analyses for cytokines and autoantibodies, and imaging studies; we would find imaging of microglial activation especially interesting.
4) Nutritional deficiencies related to microbiome issues addressed to determine if symptoms improve.

Hypothesis 3: Long-COVID Syndrome is Related to Persistent SARS-CoV-2 Virus or Virus Fragments Sequestered in Brain Cells

Background

Multiple human viruses remain sequestered in our bodies after we recover from the initial infection. Many remain in a latent state forever, but some, such as herpes simplex or herpes zoster, appear to escape from latency periodically to produce disease or damage. In some cases, vaccination to build up the immune reaction to these viruses can suppress their disease-producing reemergence.

In rare cases, one of the most infectious human viruses, the measles virus, causes a progressive, and almost always fatal, neurodegenerative disease. After the acute infection, the virus undergoes a series of mutations while residing in host neurons and oligodendroglia. These mutations prevent the formation of intact virus, so it remains sequestered. However, the disabled viral particles can infect other neurons or oligodenroglia by direct contact from cell to cell, slowly spreading the virus through cerebral tissue producing serious disease [21].

Other viruses remain in body tissues as viable viruses, long after apparent recovery from the illness, and even remain transmissible. Ebola virus, one of the deadliest viruses known to man, was found in the semen of survivors up to six months after recovery from the initial infection. Such individuals can infect sexual partners months after infection.

The status of possible residual infection of the SARS-CoV-2 virus after recovery from COVID-19 is not well known. It appears that some patients may excrete live virus for weeks in their feces after other signs of infection have disappeared, and independent of the initial infection's severity [22, 23].

Early autopsy studies of patients who succumbed to COVID-19 found little evidence for SARS-CoV-2 virus in the brain, or other tissues, even using very sensitive PCR tests. This was very surprising given the virulence of the virus, the presence of ACE-2 receptors on some neurons and other brain cells, and the severe symptoms seen in these patients in multiple organ systems. It was hypothesized that much of the organ damage from the initial serious infection was due to the exuberant immune response and that the virus had been cleared earlier in the infection.

Other studies, particularly one performed at the NIH and several affiliated institutions, found that nearly all their subjects revealed extensive evidence of SARS-CoV-2 in multiple tissues for up to 280 days after symptom onset [24]. This was based on autopsy reports of patients who succumbed to severe, and occasionally mild, cases of COVID-19. Viral signatures were also found in all areas of brain and other nervous tissue examined, without signs of extensive inflammatory lesions. This was noted even in very early cases where the individuals died from non-COVID-19 causes, one of whom had a known genetic seizure disorder and was only discovered to have SARS-CoV-2 after death from a seizure. They also "observed a paucity of inflammation or direct viral cytopathology outside of the lungs." None of these autopsy studies were performed in people specifically identified with the long-COVID syndrome but given the long interval between initial infection and autopsy, some may likely have had long-COVID symptoms.

Another area of research might involve the role of reverse transcription of SARS-CoV-2 RNA into DNA that is incorporated into the genome. The conversion of viral RNA into DNA occurs in many body cells by a process called reverse transcription. This has been postulated to occur for SARS-CoV-2 RNA in non-dividing brain cells as well as other cell types [25]. However, this observation remains a bit controversial [26] and requires further validation. It is not clear if the resulting DNA is active or is suppressed like DNA outside of a given cell's normal activation pattern. In some cases, however, transcription of RNA from the reverse-transcribed and integrated viral DNA may occur and the RNA may leak into the CSF. These RNA fragments could then be identified by PCR tests, mistaken for intact virus, and thereby producing false positive results. In the absence of this expression of RNA fragments, PCR tests for viral RNA in routine nasal swabs will not detect such integrated viral DNA, but the potential for genomic integration was not tested on CSF samples [27]. It is possible that viral RNA or translated protein fragments could be intermittently formed and secreted from brain cells and activate an immune reaction in an already sensitized immune system to exacerbate immune-mediated symptoms. Such a scenario could sustain the immune attacks that we postulate are responsible for the long-COVID syndrome.

Experiments that might support or refute the hypothesis that SARS-CoV-2 establishes a persistent infection in brain or other neurological tissues

1) Pathological examination of brain and peripheral nervous tissue for evidence of latent virus or sequestered viral fragments. This should include electron microscopy as well as immunochemistry and PCR evaluations.
2) Determine if gene expression patterns differ in subpopulations of the brain or peripheral neurons that might be involved in an initial infection by the SARS-CoV-2 virus and/or caused by a latent virus.
3) Using tissue collected at autopsy or biopsy (if possible), attempt to express latent virus in vitro. Several viral genes have been identified in the conversion from latent to lytic infections that may be part of the SARS-CoV-2 process if a similar conversion can be identified.
4) Determine if EBV can establish a latent infection in microglia and initiate a chronic activation that supports a continued autoimmune attack with the brain. It is also possible that latent SARS-CoV-2 could exist in microglial cells and initiate a chronic, activated state. It is also, more remotely, possible that SARS-CoV-2 virus could immortalize microglia as EBV does to plasma cells. This could then lead to overstimulation of the intracranial innate and adaptive immune systems postulated to be responsible for long-COVID symptoms.
5) Look for evidence of reverse-transcribed viral RNA to DNA with appropriate genetic techniques. If such is found, identify any consequence of such cellular invasion, such as changes in gene expressions in these cells or alterations in the neuronal properties of the affected cells. This would likely require animal models and cell culture rather than in human subjects.
6) Using quantitative stereological techniques, perform careful neuropathological exams of human patients who died from acute COVID-19.
   a) These exams should look for a loss of neurons in specific brain regions, especially those related to autonomic functions. Similar analyses of internal, receptor-bearing tissues, such as baroreceptors in aorta and carotid

arteries and neuroendocrine organs, should be undertaken. These studies might indicate neuronal or other cell loss not detected by routine imaging or autopsies that could correlate with specific long-COVID symptoms. Such a localized cell loss could be a direct effect of viral infection or direct cytotoxic immune attacks on subsets of neurons in the brain, as well as a possible persistent infection that induces apoptosis in vulnerable brain cells.

7) Determine if neurons known to express surface ACE-2 receptors do not express them in critical areas. ACE-2 receptors are shed or internalized under certain conditions, which could lead to increased inflammatory activity in areas where they are missing. This could occur during the acute infection, a later autoimmune attack, or as part of a longer term subtle, and hard to identify, process.
8) Determine if there are physiological effects of latent viruses in SARS-CoV-2-infected neurons or neurons harboring other latent viral infections. Such latent infections may alter their normal functions in the networks to which they belong. Post-latency effects within neurons caused by the measles virus is a model of how the cellular processes can be dramatically altered by cellular modification of the latent virus.

Possible therapeutic strategies resulting from this hypothesis:

Any evidence that long COVID is related to a latent infection in the brain or other body tissue would suggest that therapeutic trials to diminish the effect of virus in the CNS are essential. These would include trials with new anti-viral agents and should determine their ability to dampen microglial activation. Furthermore, in depth investigation of how latent viruses affect the basic physiology of the brain and other neural tissues, could lead to greater insights for other common syndromes, such as CFS and fibromyalgia, that are linked to possible autoimmune modulation but remain poorly understood.

References

1 Levin, L., Munger, K., O'Reilly, E. et al. (2010). Primary infection with Epstein–Barr Virus and risk of multiple sclerosis. *Ann. Neurol.* 67: 6.

2 Lanz, T., Brewer, R., Ho, P. et al. (2022). Clonally expanded B cells in multiple sclerosis bind EBV EBNA1 and GlialCAM. *Nature* 603: 321–327.

3 Chen, T., Song, J., Liu, H. et al. (2021). Positive Epstein–Barr virus detection in coronavirus disease 2019 (COVID-19) patients. *Nat. Sci. Rep.* 11: 10902.

4 Gold, J., Okyay, R., Licht, W. et al. (2021). Investigation of Long COVID prevalence and Its relationship to Epstein–Barr Virus reactivation. *Pathogens* 10: 6.

5 Su, Y., Yuan, D., Chen, D. et al. (2022). Multiple early factors anticipate post-acute COVID-19 sequelae. *Cell.* 185: 881–895.

6 Song, E., Bartley, C., Chow, R. et al. (2021). Divergent and self-reactive immune responses in the CNS of Covid-19 patients with neurological symptoms. *Cell. Rep. Med.* 2: 5.

7 Shen, X., Geng, R., Li, Q. et al. (2022). ACE2-independent infection of T lymphocytes by SARS-CoV-2. *Signal Trans. Targeted Ther.* 7: 83.

8 Constantino, C., Baecher-Allan, C., and Hafler, D. (2008). Human regulatory T cells and autoimmunity. *Eur. J. Immunol.* 38: 4.

9 Peluso, M., Deitchman, A., Torres, L. et al. (2021). Long-term SARS-CoV-2-specific immune and inflammatory responses in individuals recovering from COVID-19 with and without post-acute symptoms. *Cell Rep.* 36: 109518.

10 Mousten, I.V., Sorensen, N.V., Christensen, R.H.B., and Benros, M.E. (2022). Cerebrospinal fluid biomarkers in patients with unipolar depression compared with healthy individuals: a systematic review and meta-analysis. *JAMA Psychiatry.* 79: 571–581. https://doi.org/10.1001/jamapsychiatry.2022.0645.

11 Franke, C., Ferse, C., Kreye, J. et al. (2021). High frequency of cerebrospinal fluid autoantibodies in COVID-19 patients with neurological symptoms. *Brain Behav Immun* 93: 415–419.

12 Pu, H., Ma, C., Zhao, Y. et al. (2021). Intranasal delivery of interleukin-6 attenuates chronic cognitive deficits via beneficial microglial responses in experimental traumatic brain injury. *J Cereb Blood Flow Metab* 41: 2870–2886.

13 Roohi, E., Jaafari, N., and Hashemian, F. (2021). On inflammatory hypothesis of depression: what is the role of IL-6 in the middle of the chaos? *J Neuroinflammation* 18: 45.

14 Mayer, E., Knight, R., Mazmanian, S. et al. (2014). Gut microbes and the brain: paradigm shift in neuroscience. *J. Neurosci.* 34: 46.

15 Thion, M., Low, D., and Silvin, A. (2018). Microbiome influences prenatal and adult microglia in a sex-specific manner. *Cell* 172: 500–516.

16 Zheng, D., Liwinski, T., and Elinav, E. (2020). Interaction between microbiota and immunity in health and disease. *Cell Res.* 30: 492–506.

17 Liu, Q., Mak, J., Su, Q. et al. (2022). Gut microbiota dynamics in a prospective cohort of patients with post-acute COVID-19 syndrome. *Gut* 71: 54–552.

18 Silva, Y., Bernardi, A., and Frozza, L. (2020). The role of short chain fatty acids from gut microbiota in gut-brain communication. *Front. Neuroendocrinol.* 11: 25.

19 Minerbi, M. and Fitzcharles, A. (2020). Gut microbiome: pertinence in fibromyalgia. *Clin. Exp. Rheumatol.* 38 (Suppl 123): S99–S104.

20 Lupo, G., Rocchetti, G., Lucini, L. et al. (2021). Potential role of microbiome in chronic fatigue syndrome/myalgic encephalomyelitis.

21 Garg, R. (2002). Subacute sclerosing panencephalitis. *Postgrad Med J.* 78: 63–70. https://doi.org/10.1136/pmj.78.916.63.

22 Chen, Y., Chen, L., Deng, Q. et al. (2020). The presence of SARS-CoV-2 RNA in the feces of COVID-19 patients. *J. Med. Virol.* 92 (7): 833–840. https://doi.org/10.1002/jmv.25825.

23 Tejerina, F., Catalan, P., Rodriguez-Grande, C. et al. (2022). *BMC Infect Dis.* 22: 211. https://doi.org/10.1186/s12879-022-07153-4.

24 Chertow, D., Stein, S., Ramelli, S. et al. (2022). SARS-CoV-2 infection and persistence throughout the human body and brain. *Research Squar.*

25 Zhan, L., Richards, A., Barrasa, M. et al. (2021). Revers-transcribed SARS-CoV-2 RNA can integrate into the genome of cultured human cells and can be expressed in patient-derived tissues. *PNAS* 118: 21.

26 Parry, R., Gifford, R., Lytras, S. et al. (2021). No evidence of SARS-CoV-2 reverse transcription and integration as the origin of chimeric transcripts in patient tissues. *Proc Natl Acad USA* 118: e2109066118. https://doi.org/10.1073/pnas.2109066118.

27 Briggs, E., Ward, W., and Rey, S. (2021). Assessment of potential SARS-CoV-2 virus integration into human genome reveals no significant impact on RT-qPC COVID-19 testing. *PNAS* 118: 44.

Index

Unravelling Long COVID, First Edition. Don Goldenberg and Marc Dichter.

C

l

m

t